COMMUNING
DATA LITERACY

COMMUNING DATA LITERACY

TOOLS AND CONCEPTS FOR SOCIAL ENGAGEMENT

Katherine M.A. Reilly
Esteban Morales
María Julia Morales González

McGill-Queen's University Press
Montreal & Kingston • London • Chicago

ISBN 978-0-2280-2614-3 (paper)
ISBN 978-0-2280-2615-0 (ePDF)
ISBN 978-0-2280-2616-7 (ePUB)

Legal deposit fourth quarter 2025
Bibliothèque et Archives nationales
du Québec

Printed in Canada on acid-free paper that is 100% ancient-forest-free, containing 100% sustainable, recycled fibre, and processed chlorine-free.

This book has been published with the help of a grant from the Federation for the Humanities and Social Sciences, through the Awards to Scholarly Publications Program, using funds provided by the Social Sciences and Humanities Research Council of Canada. Funding was also received from the University Publications Fund, Simon Fraser University.

Funded by the Government of Canada | Financé par le gouvernement du Canada | Canada

Canada Council for the Arts | Conseil des arts du Canada

We acknowledge the support of the Canada Council for the Arts.
Nous remercions le Conseil des arts du Canada de son soutien.

McGill-Queen's University Press in Montreal is on land which long served as a site of meeting and exchange amongst Indigenous Peoples, including the Haudenosaunee and Anishinabeg nations. In Kingston it is situated on the territory of the Haudenosaunee and Anishinaabek. We acknowledge and thank the diverse Indigenous Peoples whose footsteps have marked these territories on which peoples of the world now gather.

LIBRARY AND ARCHIVES CANADA CATALOGUING IN PUBLICATION

Title: Communing data literacy : tools and concepts for social engagement / Katherine M.A. Reilly, Esteban Morales, María Julia Morales González.
Names: Reilly, Katherine M. A., 1974- author | Morales, Esteban (Lecturer in media and digital cultures), author. | Morales González, María Julia, author.
Description: Includes bibliographical references and index.
Identifiers: Canadiana (print) 20250230763 | Canadiana (ebook) 20250230895 | ISBN 9780228026143 (paper) | ISBN 9780228026167 (ePUB) | ISBN 9780228026150 (ePDF)
Subjects: LCSH: Big data—Latin America. | LCSH: Personal information management—Latin America. | LSCH: Internet—Social aspects—Latin America. | LCSH: Internet literacy—Latin America.
Classification: LCC QA76.9.B45 R45 2025 | DDC 005.7—dc23

This book was designed and typeset by pata in Exo and Minion Pro.
Copyediting by Grace Seybold.

McGill-Queen's University Press
Suite 1720, 1010 Sherbrooke St West,
Montreal, QC, H3A 2R7

Authorized safety representative in the EU:
Mare Nostrum Group BV,
Mauritskade 21D, 1091 GC Amsterdam,
the Netherlands,
gpsr@mare-nostrum.co.uk

CONTENTS

SECTION THREE

Techniques to Commune Data Literacy

TABLE, EXERCISES, AND EXAMPLES

TABLE

EXERCISES

EXAMPLES

meetings, transferring funds, and making payments. Projects of this kind truly cannot happen without administrative support, and we are deeply grateful for your efforts.

We have learned so much with and from the participants who engaged in dialogues and workshops in Colombia, Peru, Paraguay, Uruguay, and Chile. None of the ideas presented here would exist without your willingness to explore your data use. Our gratitude to all of you.

The literature reviews, field work, data collection, and analysis that formed the backbone of this volume were supported by various research assistants. Your efforts have profoundly enriched our work. We thank Nicole Stewart, Carolina Carrereto, Ramiro Liesegang, Belen Febres, Marieliv Flores, Rachel Horst, and Jihyun Park for their insights and contributions to this project.

We would also like to thank the people and spaces that inspired us and helped us improve the ideas we bring forward in this book. Thank you to Gabriel Kaplún for helping to identify Latin American communications literature. Thanks also to the Tierra Común (https://www.tierracomun.net/) community for your support and inspiration. Also, a big hug and the promise of a delicious meal to Ranjit Singh for your generous, creative, and constructive guidance of our writing process. And thanks to Dr Linnet Taylor from Tillbert University and Fundación Karisma for including Katherine in "Justicia de datos en América Latina: Retos y alternativas" (Data Justice in Latin America: Challenges and Alternatives), which offered a space to develop the analysis presented in this volume. We are also grateful to scores of conference-goers who provided critical feedback on various iterations of this work during the conferences of the Society for the Social Study of Science (4S), the Association of Internet Research (AOIR), the Latin American Studies Association (LASA), and Data Justice.

Finally, we would like to extend our gratitude to those who provided comments on early versions of our manuscript. They include Frederik Lesage, Ulises Mejias, Rachel Horst, Wayne Carrigan, two anonymous reviewers, and our editor, Khadija Coxon. You provided invaluable guidance as we refined the ideas contained in this volume.

The cover depicts a work called *The Reader* by Stanza, which "acts to reposition a future view of ourselves when embedded technologies will become part of our everyday lives." It captures the experience of digital subjectivity and projects the need for new forms of digital literacy – themes that we explore in this book.

COMMUNING
DATA LITERACY

INTRODUCTION

Data Organizes Our Social Relations

In our digitally mediated world, fan associations, members' unions, and supporter groups have real agency over the sports teams, clubs, and public figures they follow. From Taylor Swift with her Swifties to Wrexham AFC and its Supporters Trust, the size and significance of these groups is made possible by digital networks, and participation is not only easy, but actively encouraged. Swifties express their needs and desires directly to Taylor Swift through social media, and they also mobilize to advance causes that Swift herself endorses. Similarly, the Wrexham Supporters Trust is a virtual body which expresses the interests of the Club's fans to its ownership. Examples like these give us the impression that smartphones and social media apps support community connections, bring us closer together, enable dialogue, and give us influence.

However, the implications of digital media and smartphones for communities are more complex and less tangible than these examples might suggest. Digital systems used by insurance schemes, health providers, membership programs, and banking services *also* shape our experiences of community. Since these systems tend to prioritize consumerism, convenience, individualism, and seamless interactions, they often come at the expense of community connection, transparent relations, or having a voice. Meanwhile, while these spaces and services provide real benefits in the form of discounts, efficiencies, and information, these benefits come at a cost. What many people don't realize is that our data forms the basis of information systems that organize our relationships, and that our data is often used in ways that divide us from each other. Because these systems are multiple, embedded, and difficult to comprehend, they can make us feel alone in a sea of information systems, which can make it difficult to comprehend how datafication is shaping our lives.

For example, Starbucks has one of the largest loyalty programs in the world: Starbucks Rewards. Every time a member makes a purchase, they share data with Starbucks, and the company uses that data to organize its customer base around consumer concerns such as discounted purchases, the latest secret menu drink recipe, or where to score a new collector's mug. However, the use of personal data by loyalty programs is very opaque, and this makes it difficult to form groups and express power over the terms of our participation in Starbucks's loyalty scheme. It's not immediately obvious who else is a member of Starbucks Rewards, and this makes it difficult for loyalty program members to come together and share their views. Starbucks, meanwhile, only encourages community to the extent that it promotes the Starbucks brand. It certainly wouldn't encourage the formation of Starbucks data rights groups, because they might demand transparency about how personal data is used to set prices, select store locations, or manage workers, and that would affect Starbucks's bottom line.

It doesn't have to be this way. Over the past few decades many groups have placed conditions on who can access their data and what it can be used for. For example, Indigenous communities have expressed clear sovereignty over their data, and demand that uses of Indigenous data by governments, researchers, or businesses be negotiated on a case-by-case basis (Kukutai and Taylor 2016). In Canada, the First Nations Information Governance Centre established the principles of ownership, control, access, and possession (collectively known by the acronym OCAP) as foundations for the management and governance of Indigenous data. These principles have been widely influential, as when the Canadian federal government updated its Research Data Management Policy in 2022. This document stipulates that "data related to research by and with the First Nations, Métis, or Inuit whose traditional and ancestral territories are in Canada must be managed in accordance with data management principles developed and approved by these communities and on the basis of free, prior and informed consent" (Government of Canada 2021).

In another example, communities who come together around medical conditions often demand control over their health data, and a say in how it will be used by medical researchers, insurance companies, pharmaceutical companies, or business interests. These private, not-for-profit networks, known as patient-powered research networks (PPRNs), create secure and private channels of data-sharing to address questions prioritized by members of the network, and they also involve patients and their communities in network governance (Fleurence et al. 2014). Some examples of PPRNs include ArthritisPower, IBD Partners (which is run by the Crohn's and Colitis Foundation), and Tidepool for Diabetes. Each PPRN is different because they reflect the lived experiences, needs, and values of the people in the network. For example, the Patient-Led Research Collaborative is a group that came together around Long COVID. The group values rigorous and sound research methodologies while also prioritizing patient-led research, so part of their work has involved "developing a patient-centered research playbook for patient groups, research partners, and funders that establishes a new baseline of acceptable patient engagement in research" (Patient-Led Research Collaborative, n.d., para. 5).

A third example comes from service providers and small businesses that form data cooperatives as a way to improve services to their communities. For example, Open Data Manchester is working with an energy provider called Carbon-Coop to explore how pooled energy data can help customers reduce both their energy costs and their carbon footprint (Himmelsbach 2021). Similarly, the Growers Information Services Coop (GISC) pools member data to produce business insights about their farming enterprises. In one instance, members collaborated to produce insights about water use on farms in Nebraska, USA (Nebraska Association of Resources Districts 2019). Another example is provided by Megha, a women's farmer cooperative in Gujarat, India, which uses pooled data to create a credit rating system that improves the availability of financial supports to its members (Girish 2022).

In all these examples, specific demands about data use emerge in the context of clear communities based on careful analysis of people's experiences and careful observation of how data is used – both symbolically and materially – to shape systems, narrate realities, and influence outcomes for communities. These groups can assert their demands and mobilize alternative data practices because they have created spaces for dialogue with each other and formed a collective voice. Going back to the example of Canada's Federal Research Data Management Policy, consider how research data management is regulated for other-than-Indigenous groups. In that case, the policy says that data should be managed according to disciplinary standards, and with a view to making the data findable, accessible, interoperable, and reusable (or FAIR – see Wilkinson et al. 2016). There is no requirement to consult with research participants about how they want their data to be managed, which is likely because in most cases, there aren't clear communities to consult. It's not that other groups couldn't demand specific treatment where data management and data use are concerned. It's that, for the most part, they haven't!

We believe that many other groups would benefit from assessing the use of their personal data, and making clear demands about how it is collected, interpreted, and used. It has long been known, for example, that companies mobilize personal data to engage in price discrimination (Pedersen, Sadler, and Smart 2017). For instance, in July 2024 the American Federal Trade Commission launched an investigation into eight data intermediators including Mastercard, JPMorgan Chase, and Accenture to learn "how they combine computer algorithms and people's personal information to adjust pricing" (Kan 2024). Also in July 2024, an antitrust lawsuit was launched against the data-pooling software Real Page, which allows landlords to engage in price-fixing that raises rental rates by 3–7 per cent (Kaye 2024). There are also many examples of how data-intensive systems contribute to discrimination, as when a computer program called COMPAS, used by Florida judges to assess whether defendants were at risk of re-offending, was found to discriminate against Black people (Courtland 2018), or when, despite years of efforts to address discrimination within its network, AirBNB reported that its hosts continue to discriminate against Black travellers (Clemence 2022). In short, we know that personal data is used in ways that are unjust and harmful and that are shaping social relations.

However, for those who are trying to push back on things such as high prices, illegal evictions, or discriminatory practices, it can be very difficult to produce a clear picture of what's going on. Collective responses to extractive, predatory, or discriminatory data practices could result in huge gains for the communities that are affected by them. But as an individual it is very difficult to "see" these patterns, let alone to challenge them. Indeed, the very nature of datafication makes it difficult for people to make sense of how it affects them. This is partly because the very data used to promote individualized convenience and isolated transactions is the same data that reorganizes our daily experiences. And it is partly because datafication is designed around a series of practices and protocols that normalize data-sharing and make the process seamless, easy, and invisible. If we can't see the effect of these processes, then we are left in the dark, and it becomes difficult to know how to respond.

Communing Data Literacy explores the challenge of understanding data, engaging with community, and expressing collective needs and desires in the face of information systems that work against community dialogue or the realization of collective goals. Based on field research conducted with data justice groups in five Latin American countries, it outlines concepts that clarify how information systems can undermine our ability to make sense of personal data. It articulates alternative ways of thinking about and doing personal data literacy. And it offers tools to convene and share knowledge around specific instances of datafication. If you're interested in alternative ways of thinking about personal data literacy, if you want to explore data literacy as a tool for building community, or if you're looking for inspiration about how to design personal data literacy interventions in educational or community settings, then this book is for you!

BECOMING CONSCIOUS OF DATAFICATION

For more than twenty years, experts in North America and Europe have been ringing alarms about the growing use and abuse of personal data in the digital economy (Gandy 2011; Park 2013; Trepte et al. 2015). They have long signalled that corporate collection of personal data could be used to influence public opinion, control pricing in the marketplace, invade people's privacy, or (re)produce social biases. However, in 2018, when we started the work that gave rise to this book, wider publics were only just waking up to the social, economic, and political implications of the contemporary wave of datafication (van Dijck 2014).

What was once the domain of specialists was becoming more broadly recognized – that personal data is being appropriated in new ways to enable platform business processes (Srnicek 2017), digital payments, insurance and banking (Christl 2017), brand management and consumer loyalty programs (Aimé, Berger-Remy, and Laporte 2022), the administration of social services (López and Fundación Karisma 2020), and many other processes. And yet, despite growing awareness, most citizens and consumers still had little knowledge of how their personal data, and the behaviours, views, or preferences it captured, might serve the interests of digital power brokers. Nor were they aware of how new patterns of digital mediation might reproduce or restructure their experiences of social connection or social justice.

In this context, personal data literacy emerged as a novel strategy to protect, educate, and empower citizens and consumers. Through thoughtfully designed interventions, personal data literacy aimed to make people aware of how their data was being collected, how it was being used, and how they might respond. Innovative thinkers such as Rahul Bhargava and Catherine D'Ignazio (2015), Jennifer Pybus, Mark Coté, and Tobias Blanke (2015), Luci Pangrazio and Neil Selwyn (2019) Annette Markham (2019), Juliana Raffaghelli (2019), and Martin Tironi and Matias Valderrama (2021) began to articulate a body of research on personal data literacy in response to the changing corporate and policy context. Their exciting works reorganized the landscape for community-engaged scholarship by combining insights from digital humanities and critical data studies (Kitchin and Lauriault 2018; van Dijck 2014), design thinking (Costanza-Chock 2020; Escobar 2018), future-forming practice (Gergen 2014), and social science research methods (Gaddis 2018; Markham 2021a).

We have been greatly inspired by these works, but as scholars researching data and digital apparatuses in the Global South, we noticed that most of these projects were carried out in North America or Europe. We wanted to think about how datafication and data literacy might manifest in other parts of the world, particularly in Latin America. Furthermore, as community-engaged scholars working alongside colleagues and partners in the digital justice community, we wanted to support a conversation about personal data literacy in that region. Finally, we wanted to explore perspectives on literacy and critical data studies coming from Latin America and the Global South to see how they might inform alternative ways of thinking about and doing data literacy work. So we set out to explore personal data literacy in Latin America.

It hasn't always been obvious how to find this alternative path. Our first effort was a pilot intervention in Lima, Peru, which we implemented in collaboration with the Peruvian digital rights NGO HiperDerecho. In that project we showed citizen volunteers how to mobilize Peru's access to information laws to conduct citizen audits of corporate uses of their personal data (Reilly et al. 2020). Participants requested access to the personal data that companies held about them and then checked to see if their data was being administered in accordance with corporate policies and national laws. Through this work, participants gained critical data literacy, if by this we mean identifying how data is generated about us, understanding how it is being used by third parties, exploring the implications, and developing strategies and tactics to address this reality (Pangrazio and Selwyn 2023).

However, we came to realize that our work positioned personal data literacy as a response to digital transformations (Vial 2019) that we *assumed* were taking place. That is to say, we assumed a lot about how data was being mobilized by companies in Peru, and also how people experience or participate in corporate datafication. This was problematic because it wasn't clear that Peru was experiencing datafication as quickly or in the same ways as North American or European touchstones (Reilly and Flores forthcoming). We also worried that by showing people how to audit the personal data use of corporate actors we were asking them to do the oversight work of the state, according to the values and criteria established in state laws, rather than exploring their own experiences, needs, or desires or articulating and mobilizing their own values or perspectives. We concluded that we had fallen into the trap of reproducing European and North American perspectives on datafication (Milan and Treré 2019).

Through these initial explorations, we began to realize that personal data literacy interventions should start from a very different place – that they should explore how experiences of social connection are transformed as personal data is taken up in the coordination of day-to-day transactions and organization of social relations. Exploring these changes from the point of view of people's everyday experiences offers several important advantages over existing approaches to personal data literacy. It ensures that efforts to engage with and understand social change reflect people's lived reality of technological change, whatever that might be in the local historical context. It also puts the focus on how those changes (or lack thereof) manifest in people's social, political, and economic relations, which offers important correctives to technocentrism and Eurocentrism in critical data studies. Finally, by centring on social relations, data literacy interventions can enmesh legibility and literacies in critical assessments of class, race, and gender. This offers the potential to reorient social ties that have been appropriated and reformulated by the transactional nature, isolation-by-design, and lack of transparency of digitally mediated interactions. In this way, our approach differs fundamentally from critical perspectives that emphasize the reflexive capacity of individuals, and instead focuses on creating spaces where knowledge can emerge from and inhere in grounded connections.

Informed by this new approach, from 2020 to 2022 we collaborated with Derechos Digitales in Chile, Fundación Karisma in Colombia, TEDIC in Paraguay, HiperDerecho in Peru, and ObservaTIC in Uruguay to carry out a series of community-engaged design research experiments in personal data literacy. These organizations identified a personal data issue of local interest and convened a group of volunteer participants around the relevant database, information system, or data-intensive service. In an initial round of workshops our collaborators probed how people experience datafication in their daily lives. The project then designed prototypes of workshop techniques to facilitate dialogues in data literacy, which we passed on to our collaborators for testing. Our collaborators shared reports about their experiences, and these reports became a basis for designing concepts and techniques that can be applied in initiatives that endeavour to make datafied social relations tangible. Overall, this design-based approach explored how to engage people in making social change legible and worked to produce knowledge about data's social effects in, from, and through community efforts. In total, therefore, this book embraces Gergen's advice to "[u]ndertake research as a form of social action, with the words following after" (Gergen 2014, 307).

Through our collaborations, interventions, and theorizations, we have come to embrace a fundamental shift: away from data literacy as the acquisition of the competencies deemed functionally or critically relevant in the context of digital transformations, and towards data literacy as the process of *communing* around socio-technical transformations in ways that create or reinstate knowing and perceiving connections. By "communing" we mean respectfully sharing thoughts and feelings in the manner of *comunalidad*, an Indigenous concept from the Mixe and

Zapoteco communities of Mexico which prioritizes living and being in ways that produce a "we" (Martínez Luna 2015; Guerrero Osorio 2019). And we also mean engaging in an "aesthetics of effort" that counteracts the dominant ontologizing forces of datafication, and defines new channels of mediation and appropriation, in the manner of the Colombian concept *sentipensar* (Cepeda 2017). Our aim is to convene spaces in which people can contemplate experiences with datafication in dialogue with others such that technology's symbolic and ordering influences will become legible, and our newly materialized social connections will begin to enact alternative ways of knowing and being. We have come to think of this approach as *communing data literacy.*

HOW TO *THINK ABOUT* AND DO PERSONAL DATA LITERACY

In his historiography of data, Daniel Rosenberg tells us that the word "data" came into being in religious debates or in scientific pursuits as a way to capture "what is taken as true" and what is "given by the conventions of argument" (Rosenberg 2013, 20). When we are in conversation with others we rely on "what is given" so that we can move on to other things such as crafting explanations, arriving at a decision, making plans, or developing strategies. In this sense, data is not part of a foundational ontology or an axiomatic truth; rather, it is a social construct and a tool that we manufacture to help us do cognitive and social work. Rosenberg goes on to quote Archimedes as saying, "give me ground to stand upon, and I will move the world" (2013, 21). This suggests that data is produced to suit particular purposes, and those purposes will shift depending on the context, positions and agendas of different actors. Data is never just lying around, but is always the result of prior processes (Law 2004). It is not the beginning of a process, but rather a strand that we can pull on to unravel the knitting that came before.

Despite this, when people are taught about data and information systems, they are frequently introduced to the knowledge hierarchy or DIKW pyramid, which explains that Data is organized into Information which forms the basis for Knowledge that can be applied with Wisdom (Rowley 2007). This metaphor is popular presumably because it offers a justification for the creation of information systems: if we organize the flow of data in the right ways then we will have the knowledge we need to wisely manage a business or a process. But from our point of view, the main thing to learn from the DIKW framework is that information systems are designed for specific purposes.

For example, in the 1990s in the era of the open internet and neoliberal reform, the knowledge hierarchy offered a way to retain the accumulated wisdom of longtime workers while flexibilizing the labour force to suppress labour organization and wages. The newly emerging "information superhighway" enabled these shifts both conceptually by offering a way to imagine knowledge floating free from people, and materially by providing a hyperlinked structure to hold this knowledge. The data, information, knowledge, and wisdom of the worker were embedded in the organization through its information systems, and the worker themself could be let go.

More recently, we've seen capital take up data and information systems in a different way. In the wake of the 2001 dot-com stock market crash, tech companies repositioned data as a new way to capitalize on social media (O'Reilly 2005). Out of this came what Mayer-Schönberger and Cukier (2013) called "datafication," the quantification of social action through digital interventions in ways that allow for tracking, analysis, and prediction. In this formulation, information systems are mobilized to anticipate future needs, wants, or actions and to send targeted communications to specific individuals or groups. Here a different form of competitive advantage is gained from data and information systems – not cutting labour costs, but rather making improvements in customer/audience acquisition and retention. These extractive processes are designed and deployed to organize a continuous flow of value from people and communities (Haggerty and Ericson 2000; Gandy 2021).

To ensure a return on datafication, social processes and beliefs need to be reorganized around the new role of information systems in capitalism. There need to be digital transformations within existing organizations, which Vial defines as "a process wherein organizations respond to changes taking place in their environment by using digital technologies to alter their value creation processes" (2019, 119). And these transformations need to create and respond to new data-intensive institutions (datafication) and new discourses and narratives that motivate acceptance of these changes (dataism). Together these shifts produce new affordances (dataveillances) such as the ability to anticipate gaps in the marketplace and the ability to circulate targeted messages (van Dijck 2014).

From these two examples (labour flexibilization and datafication) we learn that data and information systems can "produce the ground from which to move the world." This of course has implications for how people think about datafication from the perspective of the Global South (or the global majority). Most obviously, the people who designed and engineered this "ground" have a great deal of influence over the symbolic and material resources that define it, and they are also the best positioned to leverage its advantages. Since datafication was led by tech giants centred in the Global North (or global minority) and since early efforts to regulate datafication (such as the EU's General Data Protection Regulation) also emerged in the West, datafication has decidedly imperial tendencies. These observations are captured in the literature on data colonialism (Couldry and Mejias 2019; Ricaurte 2019), which explores how the epistemic and material resources of data are mobilized by capital in ways that (re)produce global patterns of precarity and marginalization.

On the other hand, projects to produce the truths that move the world must always manifest in historical and spatial realities. What people experience as digital transformation and datafication in their daily lives will depend on how technologies, institutions, and ideologies manifest in their local context, and what the technologies afford or obviate in practice. In an immediate sense this means that we can never assume that technological change is unfolding in the same way in different places. (As we pointed out previously, we learned this the hard way during our data literacy project in Lima, Peru.) In a more general sense, it means that we should think of datafication not as "the quantification of social action through digital interventions in ways that allow for tracking, analysis and prediction" (Mayer-Schönberger and Cukier 2013) but rather as *the transformation of social relations that result from the uptake of personal data in everyday transactions* (Reilly and Flores forthcoming). It is the transformation of these social relations that most concerns us in this volume.

In exploring social transformations that result from datafication, we draw heavily on the work of the famous Latin American sociologist Jesús Martín-Barbero (1993). What we take from his vision is that we operate in a politically and economically structured regime of perception (or "sensorium" in JMB's words) and that data, information systems, and technology all intercede in the organization of that regime. The nature of the regime of perception shapes our ability to make sense of how it is structured, engage in grounded forms of mediation, appropriate its resources (S. Morales 2009), and orient our futures. What concerns us especially is the power of data and information systems to organize material processes and occupy the regime of perception in ways that undermine our ability to form knowing and perceiving collectives, communities, or cultures.

In particular, capital uses datafication to organize people into patterns of behaviour and consumption that ensure a constant flow of value, but these same processes of organization leave people isolated from each other. These data-intensive processes of assembly go on whether we are conscious of them or not, meaning that in our digitally mediated world we are part of literally hundreds of "communities of convenience," often without knowing. We introduce the notion of a community of convenience to highlight how people experience social structure as a result of data power and data assemblages. In this type of social arrangement, people are affiliated to the extent that they supply data, but this data is used to organize them in ways that leave them disconnected from each other. The symbolic power of data is of particular concern here. We leave digital traces when we access the spaces, services, or discounts provided by communities of convenience. This provides the convenors of these assemblies with the tools they need to organize our relationships and construct messaging that convinces us to accept these arrangements.

We experience these processes through very small transactional windows that we engage with individually and in isolation, as when we sign a consent form or complete a digital transaction. As a result, it is very difficult to make sense of our experiences of datafication. According to Martín-Barbero's theory of mediation, the imperial or colonial agendas of those who "produce the ground from which to move the world" are never completely realized. This is because cultural content and technological apparatuses get tangled up in the meaning-making, accommodation, and appropriation of social institutions such as the family or the church. But Martín-Barbero's arguments rest on a kind of epistemic equivalency between cultural producers and cultural consumers. In a broadcast world, writers and readers had an equal shot at spinning the meaning of an anecdote. But we're not certain this epistemic equivalency sustains when it comes to data and information systems. We worry that for data subjects, data-driven processes are often illegible and therefore unknowable, which undermines their potential to make sense of datafication, foreclosing spaces for self-determination. We develop the concept of the unhomed data subject to capture this experience of being hermeneutically or epistemologically adrift. By this we mean that data subjects often lack the grounds from which to make sense of or mediate transformations to their social relations that happen as a result of datafication.

In taking this position, we mark a departure from past assumptions about the nature of consciousness and its relationship to literacy, pressing back on Cartesian or Foucauldian visions of subjectivity, and foregrounding our fragmented engagement with our surroundings. From this vantage point, data literacy has little to do with technology, and a lot to do with social processes. To be data literate is to have the potential to enter into dialogue with others and to creatively counter the ontologizing nature of datafication (Gergen 2014). This means enacting forms of sociability that are antithetical to datafication, by communing with each other about our experiences of datafication. And it means connecting with each other to illuminate the larger processes of social organization we've become a part of.

Practically, for data literacy interventions, this means convening spaces of reflection and dialogue in which people can share everyday experiences of datafication to amass the pieces of the story that make up the whole. In this way we make datafication legible, and this legibility enables us to make sense of, mediate, and appropriate our datafied sensorium according to our values, needs, and desires. This is what we mean by communing data literacy. In practice, doing personal data literacy work involves convening members of a particular system and facilitating processes that make the logics of that system legible. It's about helping people see that their personal data is of collective importance. It involves enacting forms of sharing and dialogue that counteract the isolating, individualizing, and stultifying logics of datafication. Through these processes, people make sense of how they are being organized by these systems, and from there they make decisions about how they wish to respond, either as individuals or as groups. Ultimately, this can be the first step in coming together as a collective through which data holders have the potential to express and assert their needs, desires, or values around the design of datafied systems.

In what follows, we explore the design research approach that led us to these conclusions, as well as their implications for data literacy interventions. Section 1 discusses the difference between participatory action research and critical design research as an approach to producing knowledge about literacy interventions. We share five design experiments in data literacy and then discuss what we learned from them. These lessons establish a design imperative for the concepts and techniques that follow. Section 2 discusses how concepts inform approaches to digital literacy, and then shares five novel concepts that we feel can inform the process of *communing* data literacy. Section 3 leverages these concepts as a foundation for nine group facilitation techniques for data literacy workshops. Each technique includes an explanation of its logic, step-by-step instructions, and examples of how it might be applied.

We conclude the volume by reflecting on our initial goal, which was to develop alternative approaches to critical data literacy that reflect Latin American and Global South experiences, theories, and perspectives. We argue that critical literacy perspectives have tended to emerge from Western readings of datafication, surveillance, and privacy. In contrast, *Communing Data Literacy* bases itself in Southern postcolonial readings of extractivism and autonomy. It therefore offers a much-needed response to calls for critical data studies from the global periphery, as well as a welcome set of alternatives for people anywhere who want to counter the power of data over our experiences, communities, and relationships.

Section One

—

Design Research about Personal Data Literacy

TAKING A DESIGN RESEARCH APPROACH

Design plays a double role in data literacy interventions. First, design acts as a valence for how we experience datafication. Information systems and patterns of personal data use are designed according to certain values, in ways that make them difficult to know, and that have certain effects in society. These design choices and their effects need to be made legible so that we can "see" them, and make sense of the worlds they enable. At the same time, we argue that we can't be fully data literate until we commune with the people who we are affiliated with through the databases that bring us together. This is because you can never fully know a complex system or its effects by yourself. It's the mosaic that allows you to see how your own experiences and your social relations are changing. The act of sharing about our personal experiences helps us see, learn about, and make sense of the larger processes taking place (Kaplún 1998). In this sense, design plays a second role in data literacy interventions. It offers a methodology for probing people's experiences of datafication, exploring people's assumptions about what data is, and developing concepts and techniques that help people see, read, and understand how datafication is shaping their collective experience.

These observations are easy to make in retrospect, but design was not part of our vocabulary at the beginning of this project. Initially we used an approach that accommodated our very diverse project team, which consisted of activists, lawyers, students, social scientists, and community members spread across six countries. Each member brought their own ambitions to the initiative, but what brought us into synchronicity was the idea that we were doing participatory action research (PAR) with a focus on building the data literacy necessary for people to activate data citizenship (Carmi et al. 2020; Lameras, Arnab, and Lewis 2024) or participate in data justice (Milan and Gutiérrez 2015; Shreiner and Journell 2024; Gutiérrez 2019; Elisa Raffaghelli 2020). PAR was developed in Latin America (Fals-Borda 1987) and continues to influence the work of NGOs and researchers in the region to this day, so it offered a familiar touchstone for everyone involved in the project. PAR also offered flexibility for each actor to pursue their own locally appropriate agenda through combinations of research, community engagement, education, mobilization, and policy advocacy. Finally, at the beginning of this project, we were strongly influenced by the idea that people needed help learning how to defend themselves against privacy incursions (Trepte et al. 2015) or they needed critical literacies so they could appropriate digital technologies (Martin 2008) or respond to them in creative or purposeful ways (Pangrazio and Selwyn 2019; 2023; Masur 2020). Conducting research about participatory literacy interventions agreed with how other academics and intellectuals were framing the problem of personal data when our project launched.

Over time, however, design research crept into our thinking in ways that challenged our commitment to PAR, shaped our analysis, and ultimately structured the narrative of this volume. Design appeared in works that strongly influenced our thinking such as Costanza-Chock's *Design Justice* (2020) and Escobar's *Designs for the Pluriverse* (2018), and design research began to anchor discussions in the critical data studies and digital humanities communities (Soden et al. 2021; Neff et al. 2017; Schwan, Arndt, and Dörk 2022) as a means to address or rethink data's role in mediating social processes. Meanwhile, in education, where design processes have long been used to iteratively perfect pedagogical techniques (Armstrong, Dopp, and Welsh 2020), critical and community-engaged artist-intellectuals such as Rahul Bargava (2019) and Annette Markham (2019) were applying critical, participatory, or generative design theory to datafication in a clear challenge to educational orthodoxy. More recently an explosion of work in the field of futures studies has taken up design-led research processes (Dubberly 2018) as a means to centre criticality, creativity, and innovation in educational processes (Horst and Gladwin 2024; Horst 2023).

In some ways design is a natural corollary of PAR. In PAR, researchers and participants work together to solve a common problem that is context-specific. They engage in an iterative process of action, research, and reflection with the ambition of creating positive change. In our project, for example, we could say that the problem was a lack of data literacy, and associated gaps in data citizenship (Carmi et al. 2020) and data activism (Milan and Gutiérrez 2015), and that researchers and activists at NGOs came together in a community of inquiry to improve data literacy through designing, testing, and reflecting on a series of community interventions. In this example, design logics are clearly inherent to PAR because actions and interventions need to be designed, and a group of collaborators tests those designs before iterating possible solutions.

However, as the project unfolded, we started to question how PAR enabled or reproduced our assumptions about datafication, and our positionality as academics and activists. For one thing, once they selected an instance of datafication – a platform, a database, a loyalty program – our partners needed to locate volunteers from that information system to take part in community engagements. In four out of five cases, these participants weren't part of natural communities; they were simply individuals who had a relationship with the information system in question. For example, ObservaTIC convened a group of senior citizens who were part of Uruguay's national health records systems. These participants represented a range of socio-economic, educational, and political demographics and they gathered virtually from across the country. They hadn't participated in identifying a problem, and indeed in many cases didn't feel that there was any problem to solve. This meant that the case studies weren't organically grounded in a community or community of practice, and didn't involve community members in the identification of the problem or design of the study. In fact, the lack of organic communities from which to engage in research about datafication emerged as a key insight of the project.

Secondly, as noted in the introduction to this volume, we were already aware of a lack of empirical research about how people experience datafication in Latin America. From our pilot study, we knew that NGOs in the region often worked with assumptions drawn from European or North American experiences, and that these assumptions were amplified through international activist and donor networks, as well as global digital justice events such as RightsCon. If our projects were going to be grounded in local realities, then we needed to start from people's local, everyday experiences, and understand the challenges or opportunities of datafication from their points of view. This meant that we would need to mobilize design quite differently, not as a means to develop and test a solution, but rather as a means to carry out research that would inform locally appropriate forms of literacy intervention. This raised a question: appropriate designs for what? Were we still trying to give people the knowledge they needed to become good data citizens and data activists? Or was the goal something else, such as creating spaces where people could come together to imagine what their own goals might be?

Finally, our gathering concerns provoked a close examination of the foundations of PAR, and in particular the role of subjectivity in our work (E. Morales and Reilly 2023; Reilly and M.J. Morales 2024). Subjectivity concerns what we know, feel, or perceive from the perspective of our senses and sentience, and therefore it is linked to our consciousness, self-awareness, and personhood. Inspired by the work of Freire (1970; 1985) and others, PAR leans heavily on the subjectivity of knowers as a vehicle for social change. By including participants in action-oriented initiatives, PAR aims to help people become conscious of cultural hegemony, gain clarity about the structural causes of social and epistemic injustices (which are often considered to be twinned socio-ideological injustices), and develop solutions-oriented approaches. In this sense, it is a form of critical hermeneutic action that targets epistemic injustice (Fricker 2007) by repositioning the locus of knowledge production and centring people's standpoints and lived experiences.

However, one of the widely theorized effects of digital mediation and datafication is the fracturing of subjectivity (Dubet 1994; Martín-Barbero 2011; Smith 2012; Couldry and Hepp 2018; Sefton-Green and Pangrazio 2021). Dubet, for example, argues that in our contemporary world, the unity of our roles has been displaced by the diversity of our experiences, and as a result our subjectivity emerges and shifts in and through the administration of the various logics we encounter in the many fractured expressions of modern life. This fundamentally challenges the idea that individuals have clear standpoints that can serve as a counterpoint to unified hegemonic positions, or that dominant cultural norms are the main force holding people back from creating social change. And this calls into question the notion that data literacy can directly tap into people's roles as citizens or advocates. This suggests that different approaches are required, ones that make legible people's everyday experiences, assumptions, and practices, and that allow people to navigate experiences of fragmentation in the context of datafication and associated socio-technical processes.

Design research, and in particular critical design perspectives, offered an alternative way to move forward in the face of opacity and fragmentation while still working productively with our partners and their community-engaged initiatives. The first round of workshops carried out by our NGO partners was a focus-group–style design probe (Boehner, Gaver, and Boucher 2014; Gaver, Dunne, and Pacenti 1999) of participants' values and experiences. In design research, cultural probes are used to gather broad insights about people's culture, values, or thinking, which offers inspirations for the possible design of interventions. Based on these materials we developed prototypes for data literacy interventions. In the second workshop, the partners mobilized these techniques, which were designed to "challenge conventional values" (Dunne and Raby 2001) and "challenge what is given by society, and thus affect how the world develops" (Johannessen 2017). Meanwhile, from our interactions with NGO partners plus our extensive reading of relevant literature, we began to develop an understanding of how datafication was being thought about among Latin American intellectuals and social change agents. The insights we gained from the workshops, and from our additional observations and analysis, helped us uncover potential new avenues for thinking about and doing data literacy. This offered the basis for developing alternative concepts and techniques that have culminated in the idea of *communing* data literacy, and in this book.

Design-led research challenged our sensibilities as a group of social scientists, but it also created room to explore key insights coming out of critical data studies. A design-first approach lets go of any pretense to organizing a flow of data that will prove or disprove a certain truth, or even inform the construction of a certain narrative. However, critical design research simultaneously prioritizes a careful grounding of research results in people's lived experiences, which offer their own form of inference. Instead of focusing on data as a source of truth or legitimation, the work can be better thought of as a sustained effort to provoke scrutiny about the assumptions, values, and socio-cultural contexts that organize our understanding or actions, as well as the ways in which they sustain particular ways of being, knowing, or doing (Law 2004). Those include our own assumptions as researchers, those of our partners, that of Western theory, and those of research participants. A critical design–led research approach allowed us to map out the terrain of practices and prejudices that organize people's relationships to datafication and data-intensive processes. The disjunctures we encountered opened space to image data literacy differently, and to design an alternative set of concepts and tools for engaging with community actors.

In what follows, we share the five case studies that resulted from our collaborations with NGO partners. These were written by our partners, and while we've translated their work into English and edited it for clarity and length, we've tried to respect their narratives and avoid sweeping any incongruities under the rug. We invite you to read them as expressions of how our partners experienced this process, and what they learned about their interlocutors vis-à-vis a particular expression of datafication. At the end of this section we share the questions and design challenges that these case studies raised for us – questions we address in the rest of the book.

Eduardo Carrillo and Diana García, TEDIC

PERSONAL DATA, VALUES, AND YOUTH

Pharmaceutical Loyalty Programs in Paraguay

Paraguay's pharmaceutical industry has experienced significant changes in recent years. In 2006, Paraguay's Technical Unit for Industrial Studies (UTEPI) reported that over 90 per cent of the country's pharmaceutical outlets were independent, neighbourhood shops (UTEPI 2006) while less than 10 per cent belonged to larger pharmaceutical chains. However, by 2014 it was clear that the market was experiencing concentration and technological change as the balance between local independent stores and pharmaceutical chains shifted. In 2015, a study presented at the pharmaceutical convention Convenfarm reported that in the period from 2005 to 2014 in the cities of Asunción and Gran Asunción, consumer preference for pharmacy chains grew from 32 to 92 per cent (ABC Color 2015; see also Ultima Hora 2013, 10). Meanwhile, in 2019 the open data portal of the National Health Surveillance Directorate (DINAVISA) indicated that the percentage of independent pharmacies in Asunción had dropped to 67 per cent, with two leading chains, Punto Farma and Farmacentro, taking 14.8 and 9.5 per cent of the market respectively.

The COVID-19 pandemic appears to have accelerated the trend towards concentration and technological change in Paraguay's pharmaceutical sector as people took advantage of online purchasing to sustain social distancing. Of particular significance, in 2020 the Cartes Group conglomerate announced the acquisition of Farmacentro, which offers one of Paraguay's leading pharmaceutical loyalty programs. This was followed in July 2021 by the announcement of a new network of primary care clinics called Rapidoc (La Nación 2021). Paraguay's Investigative Directorate of the National Competition Commission (CONACOM) issued a resolution about the Cartes Group's acquisition of Farmacentro which noted that conglomerates work to find synergies between the companies that make up their holdings (CONACOM 2021). It pointed to the introduction of a new Farmacentro Affinity credit card by BASA Bank, which is a member of the Cartes Group, as an example of synergies between the holdings of the Cartes Group.

This series of events concerns TEDIC, Paraguay's leading data justice advocate, because the country has weak, fragmented, and poorly enforced personal data protection laws. At the time of this study Paraguay did not have legal requirements for informed consent during personal data collection (TEDIC 2023). When conglomerates such as the Cartes Group assemble health provision, credit card provision, and loyalty programs, they position themselves to control significant flows of personal data. This case offers a clear example of how datafication reshapes people's experience of community as they shift from local, neighbourhood-based services to ones that are heavily mediated by complex socio-technical systems under the guidance of concentrated economic forces. While people are broadly aware of these shifts in their daily lives, they perhaps aren't fully conscious of the implications for their experience of community or their ability to express agency.

Given this, in 2021 TEDIC convened two sets of workshops with young people from Asunción and Gran Asunción. They wanted to learn how personal data collection is perceived by people who belong to the loyalty programs of major pharmaceutical chains, so they targeted Paraguay's heaviest technology users: youth aged eighteen to thirty. What is their perception of the value proposition that shapes pharmaceutical loyalty programs, and how would their relationship to pharmaceutical providers differ if corporate activities reflected their personal values? Do they see any cause for concern with these programs, and what type of response do they wish to pursue, if any?

At the first set of workshops held in May 2021, TEDIC used a technique called "My Habits, My Data" (see Reilly et al. 2020) to learn about participants' experiences with loyalty programs, online purchases, and personal data collection, as well as their perceptions of data-centric practices. The workshops revealed that participants welcome the diversity of products offered by pharmacy chains, such as cosmetics or personal hygiene supplies, and that they are highly accepting of loyalty programs because of the offers and discounts these provide. As one participant shared, neighbourhood pharmacies offer what is necessary and urgent, but in chains "you can even find what you didn't know you needed." In fact, participants said that the convenience of chains motivated larger and more frequent purchases: "My consumption increased not only on the days I went to the store each month for the offers, but I saw offers one day and said, 'I'm going to take advantage and buy this now.'"

Participants were also very accepting of the need to share what they called "normal" or "basic" data when making purchases in chains, noting that "[t]o become a loyal customer you have to register at the pharmacy, they ask you for all your data and then you go to buy, give them your ID number and they give you the discount." Sharing personal data was associated with positive outcomes such as availability, speed, accessibility, economy, practicality, and efficiency, more so than with negative evaluations such as breach of privacy, fraud, data security, or misuse of data. There were some concerns, such as exposure to spam, or vulnerability from sharing location or banking information with delivery services. However, being able to link WhatsApp to an online account, receiving personalized offers based on consumer preferences, getting SMS updates about the status of deliveries, receiving extra discounts for online purchases, and being able to link their consumer account to information about prescribed medications were all seen as valuable benefits of virtual services and loyalty programs.

Participants also appreciated receiving text-based notifications about promotions and considered this an appropriate use of their personal data, even though none of them was informed about or consented to text notifications when they signed up for their loyalty programs. However, participants said they experience a greater sense of invasion when sharing data through digital mechanisms than through face-to-face purchases from pharmacy staff. They noted that kindness and good treatment by pharmacy staff are important determinants of whether they share information. They said that often they provide data *because* the person who requires it asks in a nice way and that if the person is not nice, they may be less cooperative.

There was a strong sense that companies working in healthcare are less likely to misuse people's personal data, but even so participants linked brand prestige and corporate dominance to a pharmacy's capacity to ensure personal data protection. They shared that "capacity to influence," "publicity," "fame," and "the conviction that they sell you" generate trust because they project the image of a serious business. As one person said, "They have a lot of influence, the pharmacies. I saw it as a value and I thought 'wow,' because they have influence. They tell you 'look, you have a 45% discount if you register on the web, you can't miss this opportunity.' They influence and convince. They convinced me." The results of this first workshop demonstrate how the shift from neighbourhood networks to corporate chains happens. The value that young people give to "fame" translates into a reluctance to hand over personal details to neighbourhood pharmacies, which are perceived to be "unknown" entities.

The broadly positive view of loyalty programs, online shopping, and digital communications served as a jumping-off point for discussions at the second set of workshops held in November 2021. This event addressed the value proposition that shapes corporate data collection as well as the participant's personal values where corporate use of personal data is concerned, drawing on techniques called "value versus values" and "The Value of Sharing." The goal was to raise consciousness about the gap between personal and corporate values that shapes their relationship with chain pharmacies, while also exploring the value proposition shaping data-sharing.

Using the list of values provided by the technique "value versus values," one group of participants identified honesty, commitment, and responsibility as core values that should govern personal data collected during pharmacy interactions. They felt that companies should be committed to being "responsible with the data we are giving them," and that this should include "informing us what they are going to do and what else they are not going to do." Confidentiality emerged as a concern, with one participant saying, "now that I think about it, I guess they have a record of what medicine you order, I am not sure if they keep that information, but I imagine they do. If you can access a person's name and see their history of drug purchases, you can know a lot of things about someone." This relates to the observation that companies should be honest about the terms of the value proposition: "They ask you for your information so that you have a benefit, but they do not make it clear that they are also going to benefit from sending you things, there is no honesty there."

The second group, meanwhile, identified integrity, respect, and honesty as important personal values; however, these concepts were much more closely linked to an understanding of the value proposition at work in data-sharing. For example, members of this group felt that data-sharing put them at a disadvantage because "you give something of yours that cannot be returned." Thus, integrity might involve "a cooperation agreement. I give you my data and you give me benefits. I will cooperate with you with my information, and you will provide me with benefits." Here, there was a much broader sense of what integrity might entail. Beyond honesty about the terms of data-sharing, in this case integrity referred to corporate citizenship in general: "It would help a lot for the company to demonstrate how it operates beyond products alone. How much social responsibility does it have, and how does it express this compromise not just to clients, but to people in general?"

Having established these personal values as starting points, the groups turned their attention to an evaluation of how personal data is managed in their relationship with pharmacies. When asked "Who makes decisions about your data?" the second group laughed and said, "definitely not us!" Other comments included "I don't know if we are really data decision-makers because we're asked for our data, and we have to give it to access medicine" and also "I give them my data and they are gone. Who [makes the decisions]? I don't know." They suggested that management of personal data was carried out by distant pharmacy administrators, IT personnel, or marketing departments, and indicated that once the data is handed over, there is no way of recovering it and that pharmacies can do whatever they want with it. As the terms of the value proposition between data provider and data collector began to reveal themselves, participants started to become conscious of potential inequities, such as the potential sale of personal data to third parties for market analysis. This led one participant to suggest that in the context of data-sharing with pharmacies or healthcare, this type of activity would constitute a "betrayal of trust."

Considering these discoveries, the issue of consent mechanisms was raised. None of the young people who participated in this process had experienced consent mechanisms when it came to sharing personal data, which is unsurprising given the state of Paraguay's legal framework at the time of the study. Participants felt it was fundamental to *inform* users how personal data would be used: "They should let you know what they are going to do with your data" and there should be "policies informing the purposes of data collection and requesting approval to send information." Pharmacies should explain, for example, that "[w]e have your address so that you can order delivery and so you can access discounts." However, users also expressed doubt that pharmacies would support consent mechanisms, and they also indicated that they wouldn't want constraints on personal data-sharing to affect loyalty benefits. The majority of participants perceived the personal-data-for-discounts equation, and improved access to consumer goods, to be positive and unavoidable, even as some indicated the need for better information independent of these discounts.

Responses such as these suggest that the provision of personal data has already become a normalized part of accessing medications in Paraguay. Sharing personal data has become common and is seen as non-invasive, in contrast to the past. As one participant said, "At the beginning it was shocking to me … Now it's normal, it's part of the routine, just as they say 'payment method,' they will ask you for your ID number." It seems that there is no longer much difference between a loyalty process and the individual purchase of medicines since personal data is requested as part of every interaction. Collection of "basic data" is seen as an irreversible trend by the young people who participated in the study. However, while participants emphasized the personal discretion of individuals to share personal data, they also demonstrated a lack of knowledge about how data collected via purchasing processes is used, or when data-sharing is essential to ensuring access to medicine or pharmacy services.

These findings are troubling when considered in reference to personal agency. The workshops evidently raised participants' consciousness about the terms of the value proposition when sharing data with chain pharmacies, and it also demonstrated gaps between personal and corporate values. Participants became more literate about their relationship to the digital economy as a result of these workshops. However, their acceptance of data-sharing also seems to go hand in hand with acceptance of chain pharmacies, and from there, acceptance of concentration of power in platformized industries as in the case of Farmacentro and the Cartes Group.

If the goal were to mobilize young people to demand informed consent laws in Paraguay, these concerns would be well founded. Based on the findings presented here, participants seemed unlikely to exercise this type of agency because they are broadly accepting of current data-sharing arrangements. However, when we set expectations aside, the workshops revealed alternative grounds for collective action that shifted the focus away from just data or technology. Young people clearly still place importance on human relationships and personal treatment, as demonstrated by their preference for sharing personal data with workers over computers. Meanwhile, participants linked integrity to larger commitments to corporate social responsibility. In total, there was a shift away from data and towards people. As one participant explained, "For me it would be more credible and transparent to know that all the workers have medical insurance and basic coverage. This would make the company more attractive because if they care for their workers, then they will also care for my data." Concretely, if a company cares for its workers, then by proxy it is possible to assume that it also takes care of the security and welfare of the personal data stored in its systems.

In this sense an emerging basis for collective action seemed to be linking data integrity to corporate social responsibility. Participants in this study came to realize that the fundamental value proposition is still a matter of class relations, and to the extent that data-sharing is a new way of organizing a social compromise, that compromise must still be fair for all parties involved. The suggestion seemed to be that if Paraguay is going down the road of technologically-enabled corporate concentration, then this needs to demonstrate value for citizens in a broad sense. With this in mind, agency won't express itself at the pharmacy checkout counter, but in larger spaces of social struggle.

REFERENCES

ABC Color. 2015. "92% de las compras son en farmacias de cadena" ["92% of purchases are in chain pharmacies"]. https://www.abc.com.py/edicion-impresa/suplementos/empresas-y-negocios/92-de-las-compras-son-en-farmacias-de-cadena-1363088.html.

CONACOM. 2021. "El Directorio autorizó la operación de concentración del Expediente N° 335/2021" ["The Directorship authorized the concentration transaction in Docket No. 335/2021"]. https://conacom.gov.py/el-directorio-autorizo-la-operacion-de-concentracion-del-expediente-n-335-2021.

La Nación. 2021. "La Nación / RapiDoc: una forma accesible de cuidar la salud llegó a Paraguay" ["La Nación / RapiDoc: An Accessible Form of Health Care Has Arrived in Paraguay"]. https://www.lanacion.com.py/negocios/2021/07/10/rapidoc-una-forma-accesible-de-cuidar-la-salud-llego-a-paraguay.

National Health Surveillance Directorate. 2019. "Farmacias Habilitadas" ["Enabled Pharmacies"]. Tableau Public. https://public.tableau.com/views/FARMACIASHABILITAS_072019_110919/FARMACIAHABILITADA.

National Institute of Statistics. 2011. *Industria farmacéutica y comercio de productos farmacéuticos – Situación económica de Paraguay* [*Pharmaceutical Industry and Trade of Pharmaceutical Products – Economic Situation of Paraguay*]. https://www.ine.gov.py/Publicaciones/Biblioteca/Dipticos/12%20Industria%20y%20comercio%20farmaceutico.pdf.

Pan American Health Organization, and Ministry of Public Health and Social Welfare. 2014. "Perfil Farmacéutico Nacional" ["National Pharmaceutical Profile"]. https://www.paho.org/hq/dmdocuments/2015/Farmaceutico-Perfil-de-Paraguay.pdf.

Reilly, Katherine, Belen Febres, Esteban Morales, Denisse Albornoz, Miguel Morachimo, and Marieliv Flores. 2020. "SonMisDatos: Guía Metodológica" ["It's My Data: Methodological Guide"]. HiperDerecho. https://hiperderecho.org/wp-content/uploads/2020/07/SonMisDatos.pdf.

Servín Belén, Maria. 2010. "Estudio sectorial: sector farmacéutico de Paraguay; informe final – políticas regionales de innovación en el MERCOSUR: obstáculos y oportunidades" ["Sectoral Study: Paraguay's Pharmaceutical Sector; Final Report – Regional Innovation Policies in MERCOSUR: Obstacles and Opportunities"]. https://idl-bnc-idrc.dspacedirect.org/handle/10625/45332.

TEDIC. 2023. "Por la Defensa de los Datos Personales en Paraguay" ["For the Defence of Personal Data in Paraguay"]. Tecnología & Derechos Humanos (TEDIC).

UTEPI. 2006. "Análisis del Sector Farmacéutico Nacional" ["Analysis of National Pharmaceutical Sector"]. https://docplayer.es/17585747-Analisis-del-sector-farmaceutico-nacional.html.

Maria José Lira Morán and
Daniela Salas Zuñiga, HiperDerecho

ACQUIRING CONTRACEPTIVES AT PHARMACIES

Corporeal Autonomy and Personal Data in Peru

By acquiring contraceptives of any kind, people with childbearing capacity exercise their right to choose their life plans and reaffirm their autonomy over their bodies and sexual health. However, in some sectors of Peruvian society, premarital sex, or sex for other than reproductive purposes, is considered taboo. This can be seen as a continuation of what Peruvian feminist historian Mariemma Mannarelli (2018) calls the domestication of women through *tutelaje familiar* (family mentorship). Under this system, young women are groomed to carry the honour of the entire family, and to accept that their behaviour outside of the home is a matter of public interest. We can see this reflected in the fact that Peru has some of the lowest levels of contraceptive use among women in Latin America (Rodríguez Vignoli 2017).

To complicate matters, sexuality and reproduction have become significant sites of political contestation in Peru since the emergence of the *Ni Una Menos* movement. Discussion about the right to use contraceptives is subject to silences, omissions, and negative judgments by a Peruvian society in constant tension between religious conservatism and organized movements for the promotion of gender equality (Meneses 2019). These tensions are expressed in debates about the Peruvian Ministry of Health (MINSA)'s decision to promote contraceptives under the banner of "family planning" rather than the broader category of "sexual and reproductive health." And they've come to a head in specific acts, such as the March 2023 closure of the main office of Peruvian Institute for Responsible Parenthood – INPPARES (El Comercio 2023).

In the face of these tensions, acquiring contraceptives entails moving through a complex terrain in Peru – one that women must navigate if they wish to express their corporeal autonomy. Some young women seek to avoid public scrutiny or family displeasure by finding discreet ways to exercise this corporeal autonomy. Others openly protest the denunciation of sexual pleasure, the censorship of sexual education, and the criminalization of abortion.

While corporeal autonomy has become a site of political debate in Peru, its relationship to women's personal data sovereignty has not been explored. In this section, we explore this topic in the context of pharmaceutical loyalty programs. Due to the nature of their work, pharmaceutical companies control sensitive personal information about the health and sexual behaviour of women who purchase contraceptives. This is especially so in Peru where the Intercorp conglomerate controls more than 80 per cent of the retail supply of medicines, alongside pharmaceutical laboratories, banks, shopping malls, supermarkets, restaurants, and even educational institutions such as schools and providers of higher education. Meanwhile, Peru's pharmaceutical sector has been the subject of sanctions for price-fixing (Andina 2019) and unauthorized use of personal data (MINJUS 2018), which calls into question its respect for the digital rights of customers.

How do women who purchase contraceptives from these stores experience the collection of their personal information, and how does the pharmaceutical business model, based in data capitalism, conflict with these women's autonomy and free exercise of agency? To answer these questions, this study explored the perceptions of unmarried, middle-class, educated women from Lima, Peru, aged twenty to thirty who regularly purchase oral hormonal contraceptives from private pharmacies, and are affiliated with a pharmacy loyalty program. This profile aligns strongly with the people who most commonly purchase contraceptives in Peru. Contraceptive use is concentrated in urban areas among women aged fifteen to forty-five who self-identify as white or mestizo, and whose native language is Spanish (INEI 2021). Also, 89 per cent of contraceptives are purchased in private pharmacies by women who have completed at least secondary school (Promsex 2021).

This research was carried out using Zoom-based focus groups in 2020 and 2021 with a group of thirteen volunteers. HiperDerecho mobilized these participants through its ties to feminist groups in Lima, as well as an open call in two closed social media groups: a feminist forum and a forum for university students. The first workshop used a technique called "map of my day" which was adapted from the activity "Get 'Em Thinking about Data" from the *Digital Defence Playbook* (Lewis et al. 2018, 26). Participants were asked to recall the last time they purchased contraceptives and portray this activity as a map, paying special attention to the role that technology or digital media played in this process. This activity revealed participants' habits and opened a dialogue about contraceptive acquisition. It also helped to build trust among participants. Talking about sexuality and corporeal autonomy requires a great deal of openness, and talking first about daily acts allowed participants to gradually open up about their contraceptive use.

The second workshop focused on the idea of data or information and the role that it plays in everyday acts such as going to the pharmacy. Concepts we handle regularly as social researchers, such as datafication or corporeal autonomy, are little discussed in everyday life, so it was necessary to make connections between personal data and the everyday experiences of participants. Having established this, the third workshop mobilized the techniques "value versus values" and "Overflow." These techniques were chosen because they build consciousness about misalignments between personal desires or needs and social expectations. They were leveraged to explore how personal data and digitization mediate the tension that young Peruvian women feel between corporeal autonomy and *tutelaje familiar*.

The mapping exercise used during the first workshop opened a dialogue about the relationship between participants, contraceptives, pharmacies, and technology or digital devices. Participants mapped this relationship in many ways. Some began their contraceptive journey by googling the different forms of contraceptives available, looking up testimonials from other women online, and trying to learn about possible side effects of these medications. Others documented how they travelled from their home to the pharmacy. Still others documented how they acquired contraceptives through delivery apps or how they shared the responsibility with their sexual partner.

Despite these differences, there was broad agreement that the practice of acquiring contraceptives should be discreet so as not to raise suspicions in the family, or to avoid being policed, politicized, and judged according to Lima's patriarchal paradigms. This starts at home, according to participants who said things such as "I usually go out with the excuse like 'I'm going to the corner store' and from there I go to an ATM to get cash, and then pass by the pharmacy." Another shared that "I normally take the pills out of the blister pack and keep them in my purse, so they aren't as noticeable." A third explained that if they get up early "it's like I'm following the routine of going to University, and I'm not creating any suspicion that I'm going to the doctor."

Discretion also influenced how participants get information about sexual health. They don't ask for this information at the pharmacy because it is seen as a space to transact a purchase rather than acquire medical advice. One participant shared that she selected her contraceptive based on research on internet forums and the advice of a friend: "I found a forum where they spoke well of some pills that had levonorgestrel, but a friend recommended other pills that were much more accessible, called Microgynon, and I started taking them." Another shared that "I googled 'secondary effects of pills' to see if it's just me, or if it's really the pills that are generating this." Participants consulted resources that they could access privately and on their own terms. Google stood out as a common tool for looking up the chemical components of contraceptives, as well as possible side effects. This information was then corroborated and discussed with their close circle of female friends, with whom participants shared expectations, experiences, and possible disadvantages of following this treatment.

Another key theme was logistics: "It's a lot of logistics, no? But they blame you saying, 'Why don't you take care of yourself?' It's like, you're the one putting in all the effort, but that's the context." In describing their maps, participants talked about tracking their cycle, planning a route, or going from store to store to find the pharmacy that had their brand in stock. This involved figuring out how to fit contraceptive purchases into their daily movements. For example, one participant shared that "I started at the university. I first went to the pharmacy inside the university, but they usually don't have pills, so I had to exit through the principal gates, cross Universitaria to the pharmacy on Bolivar. They always have pills there." Another participant noted that "[m]y map reflects a kind of urban segregation because I live in La Molina, so to get to any part of Lima I have to go down Javier Prado Ave." A third said, "I go to a supermarket and since they always have a pharmacy to one side, I take advantage of this opportunity to buy." Logistics also involved making plans with partners: "we take turns depending on who has time, and we work it out on WhatsApp," and "one month I went to the pharmacy two blocks from my house and the next month 'hey it's your turn' and he had them delivered to me by Rappi." These logistical manoeuvres need to be understood in the context of *tutelaje familiar* – as strategies to assert corporeal autonomy in the context of conservative taboos.

The second workshop, therefore, focused on exploring the relationship between corporeal autonomy and data collection by corporate actors. The organizers noted that many forms of digital mediation were mentioned during the first workshop, such as Rappi, WhatsApp, telephone, Google, cycle-tracking apps, and loyalty programs, and asked participants to share their perspective on the data they present when they buy contraceptives.

This prompted participants to speculate about information they might be sharing: "I don't know if I share information, but I do know that I search the internet, or ask my friends about contraceptives on WhatsApp. I guess this allows the social media people to see my search patterns?" As they identified different moments of information-sharing, such as using loyalty programs, credit cards, debit cards, or the identity cards of family members, the group began to gain consciousness of the possible links between datafication and their corporeal autonomy. One participant noted, "I just realized that when I share this information, if someone from the bank does a big data analysis, they're going to realize that first I buy condoms, then I pay for lunch, then I rent a room in a hostel." They are implying that the bank might come to realize their sexual habits and possibly also their means of earning an income. Another replied, "I never thought I was sharing my data when I searched on Google or when I paid with a card – it wasn't a worry before. Now with this focus, I'm not sure if it's good or bad to share my data."

Participants were then asked to think about the different types of data that might be collected from them. After some discussion they settled on name, ID number, forms of payment, economic capacity, sexual habits, sexual orientation, sexual health, consumer habits, and movement (geolocation), which generated discussion about control over personal data. This led one participant to exclaim: "Jeez! I feel that, no, in truth, I feel really disempowered to protect my information." Another countered by saying, "living in a digital world it's difficult to protect our data, but I think we have certain control in the sense that we can look for other ways to access ... products." A third weighed in, saying, "I feel accustomed because anyway my data is being shared not just about my reproductive health, or health, or contraceptives, but everything! To live in this virtual world, I've just accepted it." This suggested a certain level of acquiescence to the digital economy, which further comments supported: "I don't think we have much control, or there aren't many options ... You can pay with cash, but strategies like *Monedero del Ahorro* (a loyalty program) trick us into innocently sharing our information." And "Each time I click 'I accept the terms of conditions for the management of my personal data' I feel like they're killing me a little for the sake of convenience. I feel coerced because it's 'accept or get fricked.'" Finally, "We practically have no control. Each time we accept data and conditions they reduce our rights."

Having created a space for sharing during the first workshop, and having created consciousness about data-sharing in the second, the third workshop could now focus on evaluation and critical reflection. The group started by identifying personal values, which participants identified as "personal norms that guide actions" or "the basis of comportment and the correct criteria for a peaceful existence." From the list that they were given, they shared a variety of different personal values, among which were respect, honesty, compassion, equality, and humility. Participants shared, for example, that there should be a sense of equity between the person who is offering a service and the customer, and neither inequality of power nor inequality in treatment.

These values offered a touchstone for later discussion about the results of the technique "Overflow." In this case, participants were invited to share examples of moments when they were asked to provide their personal data. The exercise then asks participants to work through tensions between what they were being asked to do, how they experienced this internally, how they expressed themselves externally, and what they did. For example, one participant said that the first few times she was asked to share personal data she felt embarrassed, and surprised that they were able to call up her data so easily when they entered her

ID number. But after the first few times, she became accustomed and began to share data with ease. However, she noted that "internally I say to myself 'it's fine' even though I am questioning things. I would rather they didn't ask for my data, and that if there are discounts, they provide them through a different system, such as a coupon." Another thought about being asked to share her ID number to get a discount. She said, "I don't usually feel embarrassed about this sort of interaction, so I'm not bothered about sharing my data. I shared my data without any problem because there was a discount!" But despite this she also wished "there was more normalcy about these things, because sometimes there's tension [in the interaction] where sex is involved."

Throughout the third workshop, the issue of exchanging data for discounts came up repeatedly. Though participants became more aware of the data-discount tradeoff, they debated the possible benefits or drawbacks of this system. They reported feeling empowered by taking control of their bodies by using contraceptives, even if this meant giving up control over their personal information when buying them. This raised the question of whether women could express their agency by refusing to share personal data when making a purchase. While this idea was positively received by some, others pointed out that incentive programs are designed to dissuade people from protecting their privacy. They pointed out that women are not free to choose, because the cost of withholding data is a loss of convenience, and possibly an increased monetary cost to corporeal autonomy (through the loss of discounts).

In total, the "Overflow" process helped participants understand how they are sharing personal information by using loyalty programs and credit cards. Participants acknowledged that at some point in this purchase journey they were induced to hand over their personal information to get a discount. One said, "It's so natural at the moment of making a purchase – not even an eyebrow raised." And yet there was also the feeling that "I am literally supplying your database, so give me more!" or "I feel like it's Big Brother!" Participants expressed a desire for a better division of the benefits of that data-sharing, with themselves, and also with private companies and other institutions such as research centres or government entities. As one person noted, "What they do is analyze consumers to their own benefit, and not for the benefit of the group," and another shared, "If they are already profiting from my data, the least would be to give some benefit to us."

These observations didn't necessarily align with the participants' stated values of respect, honesty, compassion, equality, and humility. What's challenging about the link between bodily autonomy and data sovereignty in this case is the disjuncture between the consumer identity and the feminist identity. When people buy contraceptives, they seek autonomy over their reproductive health in a social context that projects gendered expectations. But pharmacies position them as consumers who are transacting a purchase. This makes it very difficult to know how a value such as equality might be operationalized. Equality might be expressed through giving all customers access to a loyalty program that provides discounts, but this same act may enable the continuation of gender norms and taboos around sexual expression.

Looking back at this project, it would have been advantageous to have participants identify their *feminist* values as a baseline for evaluating data-sharing in the context of contraceptive purchases. It would also have been helpful to know what kinds of data pharmacies gather about women who purchase contraceptives, which is possible in Peru given the existence of strong digital rights legislation. Armed with this information, it would have been possible to contrast a feminist value such as bodily autonomy with corporate databases that identify gender or consumer habits related to reproductive products. This type of analysis would allow for careful reflection about how gendered social expectations are expressed by loyalty programs.

However, the workshops were still productive. Participants identified clear associations and contradictions between what they need or want, and what is expected of them by a socio-technical system. They came to realize that pharmacies facilitate contraceptive purchases that are otherwise difficult to realize in a conservative society marked by *tutelaje familiar*. When pharmacies enable discounts, anonymous digital communications, and easy modes of delivery, women have greater control over their bodies and their sexual health.

The participants also came to realize that social pressures carry over into socio-technical spaces in ways that undermine the advancement of feminist objectives in the areas of health communication and corporeal autonomy. While health centres provide contraceptives free of charge, pharmacies offer the convenience of self-medicating without a prescription. The fact that women choose pharmacies is actually the result of failures in official information systems which position contraceptive use as a form of family planning rather than a means to manage sexual and reproductive health. These failures enable the formation of a transactional culture built on consumer values, making it more difficult to mobilize feminist values in the wider society.

REFERENCES

Andina. 2019. "Concertación de precios de medicamentos: PJ ratifica sanción de Indecopi a farmacia" ["Drug price fixing: PJ ratifies Indecopi sanction to pharmacy"]. https://www.andina.pe/agencia/noticia-concertacion-precios-medicamentos-poder-judicial-ratifica-sancion-indecopi-755894.aspx.

El Comercio. 2023. "Municipalidad de Jesús María e Inppares se enfrentan por polémico cierre de local" ["Municipality of Jesús María e Inppares Clashes over Controversial Closing of Premises"]. https://elcomercio.pe/lima/municipalidad-de-jesus-maria-clausuro-local-de-inppares-por-no-contar-con-licencia-de-funcionamiento-noticia.

Lewis, T., S.P. Gangadharan, M. Saba, and T. Petty. 2018. *Digital Defense Playbook: Community Power Tools for Reclaiming Data.* Detroit: Our Data Bodies. https://www.odbproject.org/wp-content/uploads/2019/03/ODB_DDP_HighRes_Single.pdf.

INEI. 2021. "Encuesta Demográfica y de Salud Familiar 2021" ["Demographic and Family Health Survey"]. https://proyectos.inei.gob.pe/endes/2021/INFORME_PRINCIPAL/INFORME_PRINCIPAL_ENDES_2021.pdf.

Mannarelli, María Emma. 2018. *La domesticación de las mujeres: patriarcado y género en la historia peruana* [*The Domestication of Women: Patriarchy and Gender in Peruvian History*]. La Siniestra Ensayos.

Meneses, Daniela. 2019. "Con Mis Hijos No Te Metas: un estudio de discurso y poder en un grupo de Facebook peruano opuesto a la 'ideología de género'" ["'Con Mis Hijos No Te Metas': A Study of Discourse and Power in a Peruvian Facebook Group Opposed to 'Gender Ideology'"]. *Anthropologica* 37 (42): 129–54. https://doi.org/10.18800/anthropologica.201901.00.

MINJUS. 2018. "Resolución Directorial No. 1182-2018-JUS/DGTAIPD-DPDP" ["Directorial Resolution No. 1182-2018-JUS/DGTAIPD-DPDP"]. https://cdn.www.gob.pe/uploads/document/file/1378278/RD%201182-2018.pdf.pdf.

Ombudsman's Office. 2021. "Supervisión a los servicios de atención de la salud sexual y reproductiva de las mujeres en el contexto de estado de emergencia sanitaria por el Covid-19" ["Monitoring of Women's Sexual and Reproductive Health Care Services in the Context of a State of Health Emergency by Covid-19"]. Defensoría del Pueblo, 16 March 2021. https://defensoria.gob.pe/wp-content/uploads/2021/05/Informe-Derechos-Sexuales-y-Reproductivos.pdf.

Promsex. 2021. "Variación de los costos de las píldoras anticonceptivas durante la pandemia" ["Variation in Contraceptive Pill Costs during the Pandemic"]. https://promsex.org/wp-content/uploads/2021/08/VariacionDeLosCostosDeLasPildorasAnticonceptivasDuranteLaPandemia.pdf.

Reilly, Katherine, Belen Febres, Esteban Morales, Denisse Albornoz, Miguel Morachimo, and Marieliv Flores. 2020. "SonMisDatos: Guía Metodológica" ["It's My Data: Methodological Guide"]. HiperDerecho. https://hiperderecho.org/wp-content/uploads/2020/07/SonMisDatos.pdf.

Rodríguez Vignoli, Jorge. 2017. "Fecundidad no deseada entre las adolescentes latinoamericanas: un aumento que desafía la salud sexual y reproductiva y el ejercicio de derechos" ["Unwanted Fertility among Latin American Adolescents: A Rise that Challenges Sexual and Reproductive Health and the Exercise of Rights"]. ECLAC. https://www.cepal.org/es/publicaciones/42511-fecundidad-deseada-adolescentes-latinoamericanas-un-aumento-que-desafia-la-salud.

Juan Carlos Lara and Patricio Velasco, Derechos Digitales

MIGRATION, PERSONAL DATA, AND FINANCIAL SERVICES IN CHILE

Between Vulnerability and Urgency

The number of foreigners residing in Chile grew from 1 per cent in 2006 and 2.1 per cent in 2013 to 4.3 per cent in 2017 (Fuentes and Hernando 2020, 416) following a significant influx of migrants from Colombia, Haiti, and Venezuela. For Chile's migrant population, access to banking is essential because it mediates their ability to work, pay rent, and send or receive remittances. However, as early as 2016, Lafortune and Tessada reported low bankarization among migrants to Chile (Lafortune and Tessada 2016, 16). According to the latest estimate of the Superintendency of Banks and Financial Institutions, in 2019 only 33 per cent of the migrant population held some form of banking product, in contrast to 97 per cent of citizens (SBIF 2019, 15).

Migrants to Chile face significant barriers to accessing the financial system. In Chile, a national tax ID number, the *Rol Único Tributario* or RUT, is required for a variety of transactions such as opening a bank account, making major purchases, getting a cell phone, joining a loyalty program at a store, or getting paid for freelance work. This is because Chile's Value Added Tax is levied and assessed at the point of transaction and the RUT is used to track it. Several of these transactions, including opening a basic bank account, also require a civic registry number, the *Rol Único Nacional* or RUN. Migrants face significant delays in accessing these ID numbers due to delays in visa processing (Torres 2019), as well as eighteen-month delays for ID card renewals (INCAMI 2020). This situation was further exacerbated by the COVID-19 pandemic. To make matters worse, despite official efforts to bridge gaps while migrants waited for visa processing, "official institutions themselves frequently ignore the document that certifies that a visa is in process" (Riveros 2020), thus further delaying access to ID numbers. When they lack a valid, unblocked national ID card, migrants are effectively excluded from the banking system, and since financial services are a condition of accessing other goods and services, both through financial intermediation and the assessment of credit risk, migrants face increased precarity and vulnerability.

This situation has important implications for how migrant populations in Chile relate to personal data. Chile's 1999 personal data protection law (*Law 19,628 On the Protection of Private Life*) was created to enable the flow of personal data between banks and companies (Viollier 2017) but offered few protections against algorithmic discrimination (Velasco and Viollier 2016). New privacy legislation was introduced into the Chilean Congress in 2017 but it had seen little progress as of 2022. Without legal protections, the use of personal data for banking activities creates the potential for discrimination and exclusion. To complicate matters further, as a result of challenges around visa processing and identity cards, migrants have access to a restricted range of financial and banking services. This narrows their ability to make choices about which institutions to work with, and what personal data to share and how. Finally, given their experience of displacement and precarity, migrants may face altered or increased risk from data leakage or loss.

Because this situation directly affects their daily lives, migrants in Chile are already conscious of how digital services and datafication affect their daily lives. Their challenge isn't to become conscious of how digital banking affects them, but rather to make the broader community aware of how its actions produce material precarity and marginalization.

Thus, the objective of this study was to understand how migrants relate to their data in the context of digital financial transactions with a view to carrying out advocacy on their behalf. This research was carried out in 2021 through two rounds of virtual focus groups with recent migrants to Chile. Participants were first identified by reaching out to organizations that work with migrants, and from there through recommendations using the snowball technique. In the first round of focus groups, participants were asked to talk about their experiences with data-sharing, to bring to the surface how they think about personal data and experience the datafication of their transactions. In this round there were three focus groups with six participants in each. A second round of focus groups took up the technique "The Value of Sharing" to explore the value proposition at play when migrants share personal data during financial transactions. In this case there was one focus group that brought together volunteers from across the three initial meetings. The intention here was to identify power imbalances between data-sharers and data-collectors and consider how these imbalances may contribute to the vulnerability or precarity of migrants.

In Chile, financial institutions tend to assume that foreigners have resolved their immigration status, which for practical purposes implies having a valid, current, and unblocked national ID card. This document allows anyone to open a bank account. For migrants, the most popular bank account is the Banco Estado's *CuentaRUT*, which is a basic bank account that provides access to ATM and wire services, and is available to anyone with residency in Chile. Participants pointed out, however, that accessing this account can be difficult and arbitrary. As one person explained, "I opened a Banco Estado account last year and I had an expired ID card since 2018 and they didn't give me any problems. But my husband could not open the account and his card has been expired since 2018, but his ID card is already blocked, so it was not possible with him."

Since financial institutions manage funds so that clients can receive income, pay bills, make purchases, or make transfers to third parties, they must be able to clearly identify the agents involved in a transaction. Participants understood this, saying that personal data is the information they needed to share with financial institutions to ensure the proper management of their funds. As one person said, "The truth is that I have never worried about sharing that data because they are the ones that are commonly required, for example, the account number, the account holder," and another stated, "I believe that [the data] are personal and I think it is also necessary. It is important [for the bank] to take personal data … Without that data we could not perhaps make the transaction." However, participants also had an expectation of security and stressed the need to be attentive to risks such as fraud. They said that requests to see certain information raise suspicion and doubt. As one person shared, "I would stop the transaction if they asked me for the document number or the verifier code … I wouldn't do any transaction."

Participants were particularly attuned to fraud given challenges faced by the Banco Estado during the pandemic. As mentioned, the Banco Estado's *CuentaRUT* is the most popular banking service among migrants. It is often the first account they have access to when they arrive, and many processes are realized through this account such as paying taxes or accessing vouchers. This means that the account is used in a variety of contexts, making it more vulnerable to fraud. As one participant said, it is "the most hacked platform, most attacked. Frauds are always done through the RUT account."

During the COVID-19 pandemic as transactions were increasingly handled through virtual channels, there was an uptick in banking fraud, and banks began to change how identity was verified: "last year, there were many changes in the Bank in terms of security ... Everyone has to have the third password because there were many scams." Where previously it was sufficient to present a physical ID card, it became necessary to provide additional information to verify that the account holder really was the person on the other end of the phone. Again, participants recognized that providing information was necessary to ensure the security of their accounts. As one participant said, "I feel that the companies and the government ask for this extra data because it is the only way to really verify that you are the person you say you are." However, another participant countered that "I think that more than anything it is more digital information traffic, but not that more is asked for or that there is more bureaucracy, it is more digital information traffic, but it is the same information," meaning that they weren't collecting additional data, just working to verify the client's identity.

When it came to evaluating how banks used their personal data, participants were split. On the one hand they expect transactions to be successful, and, as noted above, understand that some sharing of personal data is necessary to enable this. They evaluate data-sharing positively when it enables a necessary process, such as paying a bill or transferring money. Participants also value being able to conduct transactions digitally that could only be carried out in person in the past.

On the other hand, unwanted advertising is seen as an indication that financial institutions mishandle personal data: "Sometimes you give your email address, and you start receiving any kind of advertisement from Banco Estado, Santander [Bank], from any bank, without having requested it." They also noted that often consent to use personal data for secondary purposes appears as a condition for accessing financial services: "They ask a question about using your data, but if you don't accept them, you can't open the account." This aspect of personal data management by financial institutions was evaluated negatively by participants.

These considerations are also tied to strong concerns about data security given the above-noted rise in fraud during the pandemic. Given their concern with security, participants expressed the need for training about digital security, especially for groups with special requirements such as the elderly, offering such statements as "I believe that there is a lack of information about ongoing changes to banking transactions" and "Sorry, this is very true because, well I'm 56 years old, no? And I'm a little behind with technology and sometimes it's very hard for me to get into platforms and things. I have to go to my granddaughter and my granddaughter sometimes tells me, 'No, I'm busy, I can't.' Then I don't know what I'm going to do. There should be another way to be better with technology, with the passwords and all those things, because it's a little difficult for me." They also shared that banks should take responsibility when there is a security breach: "if there is some kind of situation like this and we are victims of a scam, they can be the ones to help solve the situation and not simply raise their hands and say that it is your problem, and we have nothing to do with it."

Based on these interventions, migrants have clear expectations about digital transactions with financial institutions. To the extent that personal data is required for the success of the transaction, management of the data should not place migrants in a defenseless position. Also, proper handling of data is seen as an indicator of reliability, integrity, and taking an interest in the well-being of individuals. Banks can demonstrate this by having more consideration for the challenges people face when they are unfamiliar with technology, and providing more support to help them understand how personal data is used to enable financial transactions.

People knowingly provide the data required to carry out financial transactions, and this in principle affirms their agency in relation to financial institutions. Migrants in particular want access to the services that financial entities provide. But the problem lies in the unequal allocation of power between the banking industry and migrants. Migrant status creates barriers to accessing financial services in the first place, and the banking services migrants can receive are of a basic nature. Once they access these services, they have little control over how banks use their personal information.

Of particular concern is how the national ID systems block migrants from accessing the financial system. In particular, several banking institutions reject the means migrants use to verify their identity during the visa renewal period. One participant shared the story of losing his bank card just as his ID card expired, which caused him to lose access to his bank account. He explained that "no paper from either police or the foreigners' office would help me," so he ended up pursuing legal action. Even after a court ruling the Bank insisted on a RUT, but it took almost two years for this to be renewed. He concluded, "I find that for migrants the treatment of Banco Estado is fundamentally discriminatory." Another participant shared that "[w]hen you run into [these banking procedures], it makes your life difficult, they truncate your dreams, your desires to move forward." In total, there was a perception among participants that they are discriminated against by banking institutions because of their migratory status.

Participants suggested procedural or legal changes that might make the system more accepting of migrants. For example, one participant wondered why Immigration couldn't issue an official letter indicating that a migrant's visa is in progress so that banks would be better able to process their requests. And in a dialogue between two participants, one person suggested that, when asked about sharing personal information, clients should be offered multiple options such as "I authorize" and "I do not authorize." The other replied, "It is a legal issue, everything would need to change upwards, because it is very complicated for us as users to make individual demands of the bank." In this sense, participants recognize the need for explicit regulations that make up for their lack of standing with financial institutions.

Increased migration to Chile happened simultaneously with the country's rapid uptake of digital financial services. The fact that access to financial services is contingent on the ability of the state to process migrants and issues ID cards creates significant barriers for foreigners residing in Chile. It causes them to face difficulties with the management of personal funds and sets up bureaucratic processes that migrants find difficult even to navigate, much less to comprehend. Resolving these challenges requires better coordination between institutions – in this case, between the banking system and the immigration offices.

Despite these challenges, once they gain access to financial services, migrants acknowledge the need to share personal data with financial institutions to process transactions. They see the value of personal data, as it is necessary to, for example, effectively connect with people in other countries and be able to send them money. Thus, even though they have doubts and concerns about how their personal data is managed by financial institutions, they resign themselves to the need to acquire a personal ID card. They also resign themselves to the options available to them in the financial marketplace. The provision of identity documentation by the authorities and the management of personal information by financial institutions constitute overlapping spheres of expectation and trust for migrants, and despite being situated outside of the system, they maintain expectations of security, respect, and service by these institutions.

REFERENCES

Fuentes, Adolfo, and Andrés Hernando. 2020. "Caracterización estadística de la inmigración en Chile" ["Statistical Characterization of Immigration in Chile"]. In *Inmigración en Chile: una mirada multidimensional*, edited by Isabel Aninat and Rodrigo Vergara, 379–408. CEP-FCE Chile.

INCAMI. 2020. "Demoras en otorgamiento de visas, permisos y trámites migratorios: INCAMI evidencia vulneración en la vida de los migrantes por este tema" ["Delays in the Granting of Visas, Permits and Immigration Procedures: INCAMI Shows that the Lives of Migrants are Violated by this Issue"]. https://incami.cl/demoras-en-otorgamiento-de-visas-permisos-y-tramites-migratorios-incami-evidencia-vulneracion-en-la-vida-de-los-migrantes-por-este-tema.

INE. 2022. "Informe de resultados de la estimación de personas extranjeras residentes en Chile al 31 de diciembre de 2021" ["Estimation of Foreign Residents in Chile as of December 31, 2021"]. https://www.ine.cl/docs/default-source/demografia-y-migracion/publicaciones-y-anuarios/migración-internacional/estimación-población-extranjera-en-chile-2018/estimación-población-extranjera-en-chile-2021-resultados.pdf.

Lafortune, Jeanne, and José Tessada. 2016. "Migrantes Latinoamericanos en

Chile: Un panorama de su integración social, económica y financiera" ["Latin American Migrants in Chile. An Overview of Their Social, Economic and Financial Integration"]. IDB. https://publications.iadb.org/publications/spanish/document/Migrantes-latinoamericanos-en-Chile-Un-panorama-de-su-inclusi%C3%B3n-social-econ%C3%B3mica-y-financiera.pdf.

Riveros, Edgardo. 2021. "Los problemas sociales que generan la demora en trámites" ["The Social Problems Generated by Delays in Procedures"]. http://radio.ucentral.cl/columnas/los-problemas-sociales-que-generan-la-demora-en-tramites.

Superintendency of Banks and Financial Institutions, Chile. 2019. "Informe de Inclusión Financiera en Chile 2019" ["Report on Financial Inclusion in Chile 2019"]. https://www.cmfchile.cl/portal/estadisticas/617/w3-article-38692.html.

Torres, Leonora. 2019. "The Integration of Migrants in Chile. Assimilation and Multiculturalist Rhetoric." *International Migrations* 10: e2068. https://doi.org/10.33679/rmi.v1i1.2068.

Viollier, Pablo. 2017. "El estado de la protección de datos personales en Chile" ["The State of Personal Data Protection in Chile"]. *Derechos Digitales*. https://www.derechosdigitales.org/wp-content/uploads/PVB-datos-int.pdf.

Viollier, Pablo, and Patricio Velasco. 2016. "Big Data: Información financiera y discriminación laboral en Chile: Un caso de estudio sobre" ["Big Data: Financial Information and Labor Discrimination in Chile, a Case Study"]. *Derechos Digitales*. https://www.derechosdigitales.org/wp-content/uploads/big-data-informe.pdf.

Ana Laura Rivoir, María Julia Morales González, Carolina Carretero, and Ramiro Liesegang, ObservaTIC

USE OF PERSONAL DATA BY PRIVATE HEALTH CARE PROVIDERS IN URUGUAY FROM THE PERSPECTIVE OF THE ELDERLY

In Uruguay, health services have historically provided universal coverage under the oversight of the State (Filgueira and Moraes 2000). In 2007, the *Frente Amplio* government pursued a series of neo-developmentalist reforms that aimed to improve service quality, enhance professionalization, and standardize logics (Larrouqué 2018; Narbondo 2015). This was realized through changes to rules for both public and private healthcare providers (Ramos and Milanesi 2016). Under these new rules, users of the National Integrated Healthcare System (Sistema Nacional Integrado de Salud – SNIS) are guaranteed access to adequate, timely, and quality health services, and are free to choose among and move between healthcare providers. As of 2019, 71.5 per cent of Uruguay's population had opted for public healthcare providers, and they also have access to private clinics.

As part of these reforms, Uruguay's government introduced a digital health records system called Mi Historia Clínica Digital that creates a unique health record for each resident of Uruguay who uses medical services. Under *Law 18.331 Protection of Personal Data*, when clinics use this database, they must guarantee the security and confidentiality of personal data, and that the contents are not unduly adulterated or shared with third parties. Also, consent to gather information must be free, prior, express, informed, and documented. The system's design is certified by the ISO/IEC 27002 standard, which is meant to ensure that only authorized agents can access health records. Health providers are responsible for establishing administrative procedures to maintain the secrecy and custody of documents (*Law No. 18.335* art. 20 2008). Finally, in the doctor's office, conversations with medical staff are confidential since the doctor-patient relationship is informed by professional secrecy (*Law No. 19.286* 2014). Clinics may also offer additional digital services for things such as appointment bookings, test results, or telemedicine. The legal provisions listed above should also apply in these cases; however, there is less standardization or oversight of digital services offered by individual clinics.

This chapter considers how elderly people perceive and make sense of this shifting situation, with a particular focus on the accumulation and use of their personal data by private healthcare providers in Uruguay. It is based on research that explored older people's concept of personal data, their data management practices, their representations and perceptions of the use and value of data in the context of healthcare provision, and their needs vis-à-vis the management of their personal data in the healthcare system. The work positions participants both as people with diverse experiences of aging, and as experts who are knowledgeable about their needs, reality, and experiences. The work explored participants' consciousness of how belonging to health information systems shaped their experience of health and aging, and discovered an interest in developing spaces where older people discuss the relationship between personal data, health, and care.

Elderly people formed a focus for this work since, while Uruguay has extensive digital infrastructure and is a regional leader according to the Information and Communications Technologies (ICT) Development Index (International Telecommunication Union 2020), the elderly still fall behind in digital uptake (Escuder 2015; Lago and Rivoir 2021). This is of particular concern given that 14 per cent of Uruguay's population is over 64 years old, according to data from the 2011 National Census. Uruguay's national strategy for digital development (Agencia de Gobierno Electrónico y Sociedad de la Información y del Conocimi 2024) included a dedicated connectivity and educational program called Plan Ibirapitá (Banco de Previsión Social 2024) that aimed to help elderly people appropriate digital technology. However, while the proportion of internet users over 65 years rose from 12 per cent to 49 per cent between 2010 and 2019 (EUTIC 2010; 2019), this is still only half of the elderly population, and two out of three of these are not daily users (Rivoir,

M.J. Morales, and Casamayou 2019; Rivoir, M.J. Morales, and Landinelli 2020). What is more, elderly people face a new context for digital uptake as personal data increasingly mediates access to essential public services such as healthcare. Uruguay's Digital Citizenship Strategy (GTCD 2020) included a series of actions aimed at increasing the competencies and skills necessary for people to enjoy well-being and exercise citizenship in digital contexts, such as understanding the social and economic impact of ICTs, knowledge of regulations, and ability to defend their rights in these spaces. However, little research has been done to inform these interventions.

To better understand elder people's perceptions of datafication in the healthcare setting, and to raise their consciousness about the implications of data collection through Mi Historia Clínica Digital, the study organized Zoom and WhatsApp workshops with sixteen people aged sixty-five and over residing anywhere in Uruguay who were affiliated with a private healthcare provider. Participants were initially mobilized through announcements sent out to the National Network of Organizations of Older Persons (REDAM), the National Organization of Associations of Retired and Pensioners of Uruguay (ONAJPU) and the Open University Non-Formal Continuing Education of Older Persons (UNI3), and then the snowball technique was used to mobilize additional respondents.

The research started with a Zoom meeting that introduced participants to each other, explained the project, and organized attendees into three WhatsApp groups according to how frequently they check their social media: every 1–4 hours, 5–9 hours, or 10+ hours. The WhatsApp groups ran for three weeks. The first week's activities brought out participants' situated views of data. An activity called "New Member of the Family" asked participants to make a list of all the things they wanted to know about a new in-law. Subsequently, they were asked to share how they might explain the concept of data to either a young person or a small business owner. During the second week, participants were asked to reflect on the traces they leave behind when they share data, how companies might use these traces, and how they feel about this. Here a version of the technique "The Value of Sharing" was mobilized, in which participants were asked to discuss the value they place on different types of data they share with their healthcare provider. The third week used the technique "Epistemic and Hermeneutic Justice" to explore what participants can know about how healthcare providers use their personal data, as well as participants' needs and desires where their personal data is concerned. Finally, the groups were brought back together for a closing Zoom session which took up the technique "Rights and Responsibilities" to explore strategies to develop data collectives and pursue further action.

While research has problematized the idea that the digital world is the exclusive domain of young people (Casado-Muñoz et al. 2015; Llorente et al. 2015; Rivoir, Escuder, and Liesegang 2019), it still tends to characterize older people as "upstarts," "digital outcasts," and "late adopters," or to frame uptake of technology in terms of the "intergenerational gap or gray gap" (Peral, Arenas, and Villarejo 2015). These characterizations are not empirically supported (Quan-Haase et al. 2018); however, they have permeated deep into the collective imaginary, homogenizing the experience of older people. This has sometimes resulted in decontextualized studies of ICT use by the elderly that prescribe attitudes, position technology as a solution to aging, or overlook the needs and desires of older people (Colombo, Aroldi, and Simone 2015; Casado-Muñoz, Lezcano, and Rodríguez-Conde 2015). For the purposes of this research, we join with scholars who reject overly generalized or paternalistic approaches to technology use among the elderly (Greenhalgh and Stones 2010; Joyce, Loe, and Diamond-Brown 2015; Neven and Peine 2017; Peine, Rollwage, and Neven 2014) and recognize multiple trajectories of old age as proposed by critical social gerontology (Hidalgo 1993; Iacub 2013).

The first workshop uncovered that participants did not have a univocal definition of data, but rather shared polysemic, situated, and relational conceptions. For them, data is a concept in dispute, where ambiguity and vagueness prevail. One participant observed that "the word is broad and can have other meanings" and then gave as an example a common phrase "*te paso el dato*," which translates literally to "I'll send you the data" but means "I'll share the details." They went on to explain that this phrase "is referring to a situation that the other person is not supposed to know about, and that possibly few people know about," and as such has a particular connotation. At the same time, participants readily recognized common types of data that might be collected in a formal setting such as name, height, documentation, nationality, date of birth, and gender.

As workshop interactions flowed, participants began to use words that appear in national regulations and circulate in international discourses such as "personal data," "public data," "private data," and "sensitive data," applying different valuations to their significance. For example, they referenced the idea of sensitive data, which according to Uruguayan law (*No. 18.331* 2008) is personal data that could reveal a person's political preferences, union affiliation, religious convictions, racial/ethnic origin, health information, or sexuality. However, they tended to introduce their own common-sense understanding into the use of these terms, engaging in what Sartori (2011) calls "conceptual stretching." For example, one participant said, "I believe all data referencing an individual is personal data that is both private and public, depending on how it gets used."

Participants recognized that digital technologies are rapidly being incorporated into various contexts, and that there is a high level of individual responsibility around their use. As has been documented elsewhere (Becker 2008; Castells 1996; Lemert 1973; Lindbeck and Snower 2001; Rivoir, M.J. Morales, and Casamayou 2019; Sibilia 2012a; 2012b), in a world mediated by big data, participants knew that if they refused to share data they would be excluded from the benefits, services, or opportunities of digital systems. As one participant shared, "of course we can refuse to give it to them, but that would imply being left out of the system. And here I don't say whether the system is good or bad, but it's the way it is … Refusing to provide data that the system asks for is to be left out of it."

Finally, participants problematized tensions between confidentiality, consent, and data-sharing. As one person said, "I think personal data is less private lately, but that's where technology and evolution has brought us." And although participants recognize that there are legal protections in place, they are unclear about what those laws cover: "I know there is a data protection law, but I don't know if it refers to all personal data. For example, I know that you can't publish the name of a person who commits a crime for the first time, but in other situations I don't know what restrictions there are."

During the WhatsApp discussion about digital traces and the value of sharing, participants indicated that when an entity, in this case a healthcare provider, is considered to be a safe space, data-sharing is not considered problematic. They said, for example, "When we are linked to a reliable medium, we don't worry about sharing our data." This is particularly so when data-sharing is seen to result in a benefit, such as improvements in health management. They also see great potential in data-sharing between different healthcare providers through Mi Historia Clínica Digital. For example, one participant said, "Well, we know that our data remains … in the computer and I am interested in that. Because in case of an emergency … they already have all our data … and they can help us. I think it is perfect that they are collecting it." Others highlighted the potential for new services such as health apps, telephone consultations, or automatic prescription renewals, as well as improvements in the efficiency or efficacy of healthcare delivery.

Despite recognizing important benefits to data-sharing, participants also raised questions and concerns. For example, they wondered how data is hosted, for how long, and the potential for its access by third parties. They noted that private healthcare providers also offer digital health platforms, and it is unclear what data these other services gather. As one participant shared, "I have no idea how data is managed by healthcare providers," and another said, "I have an app from my provider … but I don't have access to all the data they have on me."

They also noted that there is an important power imbalance between medical staff and patients. As one participant put it, "It is evident that our healthcare provider has a lot of our data that is growing over time … If we reflect critically on it, we'll go crazy because they have our life in their hands." They also noted that an exhaustive digital record does not guarantee quality medical attention, and stressed the need for healthcare that addresses the whole person.

Another concern revolved around the possibility of data leaks. Participants noted that they often receive health-related advertising and phone calls, but don't know how their contact information is being obtained. They worried about a kind of "cartel" of data-sharing between companies in the same sector: "I think it has a lot to do with healthcare providers in that data is made public for health-related companies and even private funeral homes and cemeteries. Only very specific data is kept private." And they noted that this type of activity has the potential to affect vulnerable people, including the elderly.

As for confidentiality clauses, control mechanisms, and regulation, participants felt that they matter little in the face of the human factor.

One participant reasoned that "[t]here may be various ways for data to leak from one company to the other, but it's most likely that officials from one company move to another and carry data records with them." They also believe that health providers must be sharing databases with actors in the wider health services sector to drive personalization in advertising. One participant reasoned that just as in the financial services sector, where people with a good credit record get solicitations about additional sources of credit, there must also be data-sharing in the health services sector: "in the health world exactly the same thing happens, they know that for example you are a person who has chronic diseases because in your file, in your medical history it says how many times you have been hospitalized."

In total, the elderly recognize that the flow of personal data tends to be the norm and that data has both economic and symbolic value. The existence of a single repository of health information is positively valued for its potential to improve health services, but the technology lacks utility if it is not consulted and only functions as a filing system. Finally, participants felt that there is data leakage within the healthcare system and expressed concern that this is happening without their knowledge or consent.

The discussion about epistemic justice revealed that the elderly are adapting to electronic health records, the use of applications and platforms for booking appointments and getting test results, telemedicine, and other forms of digital healthcare using multiple strategies. However, participants expressed various needs that they feel are not currently being met. These included better information about the offerings and differential costs of different healthcare providers (even though https://atuservicio.msp.gub.uy provides this information), alternatives to telephone consultations, and greater consciousness of data reporting, surveillance, and monitoring.

With regards to the latter, participants highlighted the importance of exploring, knowing, and promoting controls on healthcare providers related to the handling of their personal information. They were particularly interested in guaranteeing confidentiality in the management of their data, and one participant pointed out the importance of consent mechanisms, especially with regards to data-sharing between different health providers. They shared that their data "should only belong to the medical provider" and that they did not want it to be available to other medical service providers such as nurses, hospital transfer services, or paramedics. This concern extended to the use of health data for research: "if [our data] are going to be used for research, as a colleague has suggested, it should be done with our consent." Another argued that the need for consent and authorization should extend to national studies carried out by the Ministry of Health. Participants also expressed the need for data to be traceable so that patients could know where it was stored and who was using it for what: "If I authorize it to be used for public health or for whatever, or for research, if I authorize it, I think it is great for the good of all, but at least that I should be informed of where my data is going."

In addition, participants shared their concern about the potential for information systems to automate or depersonalize healthcare. As in the work of Sibilia (2012a; 2012b), they emphasized the need for human treatment, humanized spaces, a return to the encounter, and communion through active listening. As one participant put it, when data is used "with consistency, empathy, secrecy, delicacy, in that case I think I could be calm." Another added that data "should be handled in agreement with the user." A third participant shared "[h]ow to humanize medicine … To give the healthcare provider a social context where all those involved can feel empathy with those of us who are users. That seems to me to be a very important change, I have no idea how it could be carried out, but looking at it as a user, it seems to me that the human part is the part that needs to be modified."

These observations are closely tied to the participants' view of aging. They recognize that there are multiple ways of living and being older – that aging is part of life, does not constitute a disease, and is unique to each individual. As one person said, "The old age thing ends me. I say I am an old adult, not an old sick person. I will be ninety years old, but I can be as sick as my grandson can be with a flu, but I am not sick, I am an old adult, I am old, nothing more. So, don't make me out to [have] a disease I don't have." In this sense, the collection of health data or its analysis should be done in ways that recognize people's unique experiences with aging and enable greater connection to people seeking healthcare.

The final stage of the workshops brought people back together on Zoom to reflect on how they could better mobilize their rights and responsibilities around personal data use in the health sector. On this subject, participants reported a certain carelessness or lightness in the exercise of their citizenship both in the spaces linked to their healthcare providers and in other platformized spaces where data are exchanged. As one person put it, "I think there is also a user issue, of us citizens, who do not exercise citizenship as much as we should." Another expressed, "The issue is that users are also a bit irresponsible or complacent because we often do not demand quality of care from providers … So, I think there is a little bit of that too … Uruguayans in general, we are very passive, we accept things as they are, we do not demand, and we do not inform ourselves."

They also noted that when they do recognize that their rights are being violated, lines of complaint are not always clear, which aligns with previous findings from Waisbord (2013). In fact, in one of the WhatsApp groups, more information was requested on the regulatory scopes of data transfer, storage, and management by private healthcare entities. While participants desired a culture of consent with effective transparency around data management procedures and guaranteed confidentiality, they found that the system provides few opportunities to express demands, and that their lack of consciousness about how the system functions leaves them with little capacity to exercise their rights. This produces a kind of lean agency in which few people take responsibility for demanding the changes that might fulfill their needs and desires around use of personal data.

Emphasizing that there are few compliance mechanisms in place for data management by healthcare providers, participants expressed a need for greater oversight of data use in the healthcare sector. While participants again stressed the many potential benefits of centralized health records, one person stressed that, "like everything … if there are rules, but no possibility to control them, it is the same as if there were none, practically." Another suggested that this might be done through "some mechanism of the company itself, to ensure that these data remain in accordance with medical ethics, in silence, in the company."

As the sessions wrapped up, the participants expressed a desire to continue the conversation, but on their own terms. One said, "this group should continue because for me it has been a learning experience and each time you speak, things come up that make us question more, make us question more whether we are giving the right data or not, whether we are doing things right or not for our benefit, that's all, nothing more."

There is a need for peer-to-peer processes for data literacy that allow people to come together around their unique perspectives and concerns. These spaces could be convened by groups such as the National Network of Organizations for the Elderly, the National Organization of Associations of Retirees and Pensioners of Uruguay, or the Open University for Informal Ongoing Education of Older People. Spaces for building consciousness about data should be a continuation of earlier digital literacy initiatives such as Plan Ibirapitá, but they should focus on building critical literacy skills that help people navigate the "programmable architecture" of the platform economy. Older people need spaces to discuss and figure out how to address the implications of platformization for changing standards of care and how this is shaping their subjective experience as aging adults.

Using the data literacy techniques explored here, participants were able to gain greater consciousness of how data use in Uruguay's national healthcare system may affect their well-being and their experience of aging. By working from the experiences and knowledge of participants, and building up their understanding of what data is, how it is used, and how its use shaped their experience of aging and well-being, group members entered into a discussion about rights and responsibility with a willingness to contemplate the need for collective dialogue about data use in the healthcare system. The process allowed for the formation of critical views about data, and participants came to understand the scope and restrictions of Uruguay's new Electronic Health Record system.

The research reiterated that access is associated not only with connectivity and owning a device, but also with having the skills necessary to critically assess data requests, decide when it is acceptable to share personal information, and voice concerns or express demands when things are not as they should be. It is necessary for people to have critical skills to make decisions regarding data privacy and personal data protection, and, therefore, to know their rights and the regulatory framework that protects them. The elderly in this case study discovered that they were unaware of the protection frameworks and of the specific regulatory scopes of transfer, storage, and management of their personal data by private health entities.

While participants positively valued the fact that health institutions have their data, and that it exists in a unified form, they also recognized that they do not use it or that they underuse it personally. While they recognized that applications and systems greatly facilitate management, especially the reduction of time, they were not as aware of how and for what purpose their data are used. In fact, they acknowledged that it is used without their consent and identified circumstances in which it is used for predictions of their behaviour as a result of algorithmic intelligence. They also claimed that they benefit from this handling of their data, and while some demanded more regulation, others were willing to maintain this data release in exchange for the benefits it brings them.

In addition to the criteria of protection and confidentiality, older people demand ethical, responsible, and complete management of their data. However, the complexity that exists for the understanding of the datafication implies the need for a critical, creative, and informed attitude on the subject that involves permanent tensions between the benefits to be gained and giving up personal information in a consensual way. This complexity hinders the collective action of data collectives and the broader configuration of data citizenship.

REFERENCES

Agencia de Gobierno Electrónico y Sociedad de la Información y del Conocimiento. 2024. "Agenda Uruguay Digital" ["Digital Uruguay Agenda"]. Accessed 15 December 2024. https://www.gub.uy/agencia-gobierno-electronico-sociedad-informacion-conocimiento/politicas-y-gestion/programas/agenda-digital-del-uruguay.

Banco de Previsión Social. 2024. "Programa Ibirapitá" [Ibirapitá Program"]. Accessed 15 December 2024. https://ibirapita.org.uy.

Becker, Howard Saul. 2008. "Outsiders: estudos de sociologia do desvio" ["Outsiders: Studies in the Sociology of Deviance"]. Editora Schwarcz-Companhia das Letras.

Casado-Muñoz, Raquel, Fernando Lezcano-Barbero, and María-José Rodríguez-Conde. 2015. "Active Aging and Access to Technology: An Evolving Empirical Study." *Comunicar* 23 (45): 37–46. https://doi.org/10.3916/C45-2015-04.

Castells, Manuel. 1996. *The Rise of The Network Society*, volume 1, *The Information Age: Economy, Society and Culture*. Oxford, UK, and Malden, MA: Wiley-Blackwell.

Colombo, Fausto, Piermarco Aroldi, and Simone Carlo. 2015. "New Elders, Old Divides: ICTS, Inequalities and Well Being amongst Young Elderly Italians." *Comunicar* 23 (45): 47–55. https://doi.org/10.3916/C45-2015-05.
Grupo de Trabajo de Ciudadanía Digital (GTCD). 2020. *Estrategia de Ciudadanía Digital para una Sociedad de la Información y el Conocimiento. Uruguay 2020–2023* [*Digital Citizenship Strategy for an Information and Knowledge Society. Uruguay 2020–2023*]. Agencia de Gobierno Electrónico y Sociedad de la Inforamción y del Conocimiento. https://www.gub.uy/agencia-gobierno-electronico-sociedad-informacion-conocimiento/comunicacion/publicaciones/estrategia-ciudadania-digital-para-sociedad-informacion-conocimiento.
Escuder, Santiago. 2015. "Los múltiples escenarios de la brecha digital : perfiles de internautas según la encuesta de usos de tecnologías de la información y la comunicación del año 2010" ["The Multiple Scenarios of the Digital Divide: Internet User Profiles According to the 2010 Survey of Uses of Information and Communication Technologies"]. MA thesis. University of the Republic, Uruguay. https://silo.uy/vufind/Record/COLIBRI_8ae06636c70c4867467b194659b1bda8?sid=208671EUTIC. 2010.
"Encuesta de Uso de Tecnologías de la Información y la Comunicación 2010" ["Survey of Uses of Information and Communication Technologies"]. Agencia de Goberino Electrónico y Sociedad de la Información y del Conocimiento. https://www.gub.uy/agencia-gobierno-electronico-sociedad-informacion-conocimiento/datos-y-estadisticas/estadisticas/encuesta-uso-tecnologias-informacion-comunicacion-2010.
EUTIC. 2019. "Encuesta de Uso de Tecnologías de la Información y la Comunicación 2019" ["Survey of Uses of Information and Communication Technologies"]. Agencia de Goberino Electrónico y Sociedad de la Información y del Conocimiento. https://www.gub.uy/agencia-gobierno-electronico-sociedad-informacion-conocimiento/datos-y-estadisticas/estadisticas/encuesta-uso-tecnologias-informacion-comunicacion-2019.
Filgueira, Fernando, and Juan Andres Moraes. 2000. "Contextos y estrategias de las reformas institucionales en la Seguridad Social la Educación y la Salud en Uruguay" ["Contexts and Strategies of Institutional Reforms in Social Security, Education and Health in Uruguay"]. *Revista Uruguaya de Ciencia Política* 12: 97–122.
Hidalgo González, Jorge. 1993. "Hacia una Gerentologia Social Crítica" ["Toward a Critical Social Gerontology"]. *Revista Reflexiones* 8 (1): 1.
Greenhalgh, Trisha, and Rob Stones. 2010. "Theorising Big IT Programmes in Healthcare: Strong Structuration Theory Meets Actor-Network Theory."

Social Science & Medicine 70 (9): 1285–94. https://doi.org/10.1016/j.socscimed.2009.12.034.

International Telecommunication Union. 2020. "Índice de Desarrollo de las TIC 2020" ["Index of ICT Development 2020"]. https://www.itu.int/en/ITU-D/Statistics/Documents/events/egh2020/IDI2020_BackgroundDocument_S.pdf.

Joyce, Kelly, Meika Loe, and Lauren Diamond-Brown. 2015. "Science, Technology and Aging." In *Routledge Handbook of Cultural Gerontology*, edited by Julia Twigg and Wendy Martin, 157–65. Routledge.

Iacub, Ricardo. 2013. "Nuevas reflexiones sobre la Posgerontología" ["New Reflections on the Posgerontology"]. *Revista Kairós Gerontologia* 16 (4), "Dossiê Gerontologia Social": 295–311. https://revistas.pucsp.br/index.php/kairos/article/view/19729.

Lago Martínez, Silvia, and Ana Laura Rivoir. 2021. "Tecnologías digitales y pandemia" ["Digital Technologies and Pandemic"]. *Controversias y Concurrencias Latinoamericanas* 12 (22): 11–15.

Larrouqué, Damien. 2018. "¿Qué se entiende por 'Estado neoweberiano'? Aportes para el debate teórico en perspectiva latinoamericana" ["What Is a 'New Weberian State'? Contributions to a Latin American Perspective on a Theoretical Debate"]. CLAD *Reforma y Democracia* 70: 7–30.

Lemert, Edwin M. 1974. "Beyond Mead: The Societal Reaction to Deviance." *Social Problems* 21 (4): 457–68. https://doi.org/10.2307/799985.

Law No. 18.331. 2008. "Ley de Protección de Datos Personales" ["Personal Data Protection Law"]. https://www.impo.com.uy/bases/leyes/18331-2008.

Law No. 18.335. 2008. "Derechos y Obligaciones de Pacientes y Usuarios de los Servicios de Salud" ["Rights and Obligations of Patients and Users of Health Services"]. https://www.impo.com.uy/bases/leyes/18335-2008.

Law No. 19.286. 2014. "Aprobación del Código de Ética Médica" ["Approval of the Code of Medical Ethics"]. https://www.impo.com.uy/bases/leyes/19286-2014.

Lindbeck, Assar, and Dennis J. Snower. 2001. "Insiders versus Outsiders." *Journal of Economic Perspectives* 15 (1): 165–88. https://doi.org/10.1257/jep.15.1.165.

Llorente, Carmen, Mónica Viñaras, and Maria Sánchez. 2015. "Internet and the Elderly: Enhancing Active Ageing." *Comunicar* 45 (23): 29–36.

Narbondo, Pedro. 2015. "Las tentativas y proyectos de reforma del sistema de gestión de los recursos humanos en Uruguay: ¿nueva gestión pública o neoweberianismo?" ["Attempts and Projects to Reform the Human Resources Management System in Uruguay: New Public Management or Neo-Weberianism?"]. CLAD *Reforma y Democracia* 61: 177–208.

Neven, Louis, and Alexander Peine. 2017. "From Triple Win to Triple Sin: How a Problematic Future Discourse Is Shaping the Way People Age with Technology." *Societies* 7 (3): 26. https://doi.org/10.3390/soc7030026.

Peine, Alexander, Ingo Rollwagen, and Louis Neven. 2014. "The Rise of the 'Innosumer' – Rethinking Older Technology Users." *Technological Forecasting and Social Change* 82 (February): 199–214. https://doi.org/10.1016/j.techfore.2013.06.013.

Peral-Peral, Begoña, Jorge Arenas-Gaitán, and Ángel-Francisco Villarejo-Ramos. 2015. "From Digital Divide to Psycho-Digital Divide: Elders and Online Social Networks." *Comunicar* 23 (45): 57–64. https://doi.org/10.3916/C45-2015-06.

Quan-Haase, Anabel, Carly Williams, Maria Kicevski, Isioma Elueze, and Barry Wellman. 2018. "Dividing the Grey Divide: Deconstructing Myths About Older Adults' Online Activities, Skills, and Attitudes." *American Behavioral Scientist* 62 (9): 1207–28. https://doi.org/10.1177/0002764218777572.

Ramos, Conrado, and Alejandro Milanesi. 2020. "The Neo-Weberian State and the Neodevelopmentalist Strategies in Latin America: The Case of Uruguay." *International Review of Administrative Sciences* 86 (2): 261–77. https://doi.org/10.1177/0020852318763525.

Rivoir, Ana, Maria Julia Morales, and Javier Landinelli. 2020. "Personas mayores y tecnologías de la información y la comunicación. Evolución del uso, la apropiación y el aprovechamiento en un contexto de creciente digitalización" ["Older People and Information and Communication Technologies. Evolution of Use, Appropriation and Exploitation in a Context of Increasing Digitization"]. In *Miradas interdisciplinarias sobre envejecimiento y vejez: Aportes del Centro Interdisciplinario de Envejecimiento* [*Interdisciplinary Perspectives on Aging and Old Age: Contributions of the Interdisciplinary Center on Aging*], edited by Centro Interdisciplinario Envejecimiento, 103–27. Universidad de la República.

Rivoir, Ana, Maria Julia Morales, and Adriana Casamayou. 2019. "Usos y percepciones de las tecnologías digitales en personas mayores. Limitaciones y beneficios para su calidad de vida" ["Uses and Perceptions of Digital Technologies among Elderly People. Limits and Benefits for Quality of Life"]. *Revista Austral de Ciencias Sociales*, no. 36: 295–313. https://doi.org/10.4206/rev.austral.cienc.soc.2019.n36-15.

Rivoir, Ana Laura, Santiago Escuder, and Ramiro Liesegang. 2019. "Usos y competencias digitales en personas mayores beneficiarias de un plan de inclusión digital en Uruguay" ["Digital Uses and Competencies in Elderly

Beneficiaries of a Digital Inclusion Plan in Uruguay"]. *Revista Psicología, Sociedad y Conocimiento* 10 (1): 54–80.

Sartori, Giovanni. 2011. *Logica, metodo e linguaggio nelle scienze sociali* [*Logic, method and langauge in the social sciences*]. Bologna: Il Mulino.

Sibilia, Paula. 2012a. *El hombre postorgánico: cuerpo, subjetividad y tecnologías digitales* [*Post-Organic Man: Body, Subjectivity and Digital Technologies*]. Fondo de cultura económica.

Sibilia, Paula. 2012b. *¿Redes O Paredes? La Escuela En Tiempos De Dispersión* [*Networks or Walls? The School in Times of Dispersion*]. Tinta fresca.

Waisbord, Silvio. 2013. *Vox populista. Medios, periodismo, democracia* [*Vox Populista: Media, Journalism and Democracy*]. Gedisa.

Juan Diego Castañeda and Laura Castro, Fundación Karisma

"DO WE DELIVER WITH LOVE?"

Data from the Perspective of Colombian Gig Workers

The gig economy has had a growing presence and impact in Latin America over the past several years, which has had significant implicaions for social security (Arbeláez, Fernández, and Hernández 2021). A clear illustration is Rappi, one of the leading delivery platforms in the region. This Colombian multinational company, founded in 2015, offers an app that coordinates pickup and delivery services. Since its establishment it has exceeded a valuation of $1 billion and has expanded into nine Latin American countries. The company is gradually reaching into new areas of service provision as it seeks to prevent its users from migrating to other applications (Sanchez 2020).

The exponential growth of Rappi, both within Colombia and across Latin America, has raised issues related to job security, data use, algorithmic management, and the privacy and data protection of users and workers. Collectives of gig workers in Colombia and beyond are starting to organize to demand fair treatment from apps such as Rappi. Since August 2020, Rappi's workers, known as RappiTenderos, have come together as a union to demand regulations that ensure better working conditions and fair payments from the Colombian Ministry of Labor (Vita Mesa 2020). Nonetheless, Rappi has increasingly moved towards the automation and datafication of workflows through, for example, the implementation of a points system that determines how, when, and which RappiTenderos can work. This management through algorithms empowers Rappi to surveil and control its workers (Sánchez Vargas and Maldonado Castañeda 2020; Negri 2020), which, according to the union, increases the subordination of RappiTenderos.

Concerned about this situation, Fundación Karisma worked with eight RappiTenderos – four men and four women – to learn more about Rappi workers' interactions on the platform. Of these people, five lived in the city of Cali, one in Bogota, one in Cartagena, and one in Barranquilla. Four were permanently blocked from the platform, while the other four were active at the time of the investigation. To contact these workers, Karisma collaborated with the Platform Workers Union of Colombia, known as UNIDAPP, which represents Rappi workers. This union was formed in October 2020 to denounce labour precarity and abuses and blockades by delivery platforms.

Karisma conducted eight semi-structured interviews to learn about each person's history with Rappi, interactions with the platform, data problems, micro-resistance, needs, values, and expectations, as well as their views on the digital platformization of work. Subsequently the group members came together to share their experiences and views with each other. Based on the information collected during these first two phases, Karisma led the group in a walkthrough of the data collected by Rappi when someone makes an order through the app. This activity was similar to "The Value of Sharing" and also "Imagining Data Journeys." During this exercise, participants talked about the problems they encountered in their work, and also discussed how the app could be designed differently to better reflect their needs, values, and expectations. To conclude, the group also completed the "value versus values" exercise as a way to think through their demands for Rappi.

Among the many things that inform workers' interactions with Rappi, their main concern is how the app uses personal data for surveillance and control. When signing up to become RappiTenderos, workers agree to enable geolocation on their cell phones at all times, even when the app is closed or not in use. This affects Rappi workers whether they are working or not. As one participant explained: "[Rappi] don't have to track me with GPS. For example, if I am not working and I went to Spain, they know that I am in Spain, that is, they know where I am, so I would like to tell them 'come on, while I am not working, do me a favour and respect my space, cut the application while I am not working.'"

Participants also noted that the application violates their privacy and dignity by sending constant notifications outside working hours, when they are offline, or even if they are blocked by Rappi. As one participant shared, "Constant notifications can cause psychological illnesses such as post-traumatic stress, because they are constant. This is a form of subordination so that the worker is constantly attached to the application which only thinks about being able to generate income ... We already know that who controls this is an algorithm. By controlling us – an algorithm, which is a robot – makes us have the same action of looking like robots, to do everything they think of through this platform."

Participants also talked about the worst problem they face, which is when Rappi blocks them from using the platform. Blocking makes it impossible for workers to connect to the platform, which immediately affects their income. RappiTenderos argue that Rappi uses blocking to control and punish drivers who commit faults according to the terms and conditions of the application. It can be temporary or permanent and depends on the severity of the alleged fault. The participants in this investigation all shared that Rappi was their only source of income and that they had no other work, so being blocked represented a complete loss of employment.

Rappi blocks workers unilaterally and without giving them a chance to defend themselves. Thus, from the workers' point of view, the data gathered by Rappi is evaluated in a one-sided way, and is used against them. Rappi bases their decisions on feedback from customers or partners (such as restaurants or supermarkets), and this is the only information they take into consideration when deciding whether to block a worker. The worker's version of events is not taken into account, and this means that workers' data is undervalued by the system. As a result, participants did not consider Rappi's ratings system to be fair or proportional, but rather saw it as part of an asymmetrical data-power relationship. In addition, the reasons for blocking workers are not clearly or sufficiently explained. Participants said that Rappi's use and management of blocking "generates an insecurity or uneasiness among workers, not knowing if suddenly tomorrow he/she will be able to work. So, it is quite complicated because it is a pressure." This creates a generalized sense of mistrust among workers towards the platform.

Under the threat of being blocked, workers often feel compelled to complete deliveries that make little sense. For example, Rappi runs promotions called "RappiRecompensas" that reward workers for completing certain milestones. However, the algorithm sometimes establishes unreal expectations for the completion of these tasks. As one participant explained, "Every so often, the app offers 'RappiRecompensas.' They have these promotions of placing 7 orders for 20,000 pesos but they do not realize that they send you to place them over long distances, [it] takes hours and that you have about 3 minutes to deliver to the client."

Participants also reported being sent to unsafe areas or non-existent addresses, or arriving at a location only to discover that the platform would not let them finish the order. As one participant said, "Today I woke up blocked, most likely it is because of an order that I did not carry because the area is dangerous. I pointed it out in the Support chat and having no response I decided to free myself [of the order]. Today I woke up blocked. Do I have to risk my life and my bike to work with you?"

These experiences are further complicated by a lack of clarity about how Rappi's algorithms function. Workers say that their work and behaviour are organized by the platform's algorithms on the basis of data that is collected from and about them. They learn to work within the algorithm's conditions, so when the platform decides to make changes it creates uncertainty for workers. They are forced to adapt, which is difficult since they don't have access to the same information as the platform (Athreya 2020; Choudary 2018; Mateescu and Nguyen 2019). The need to adapt to daily or hourly changes in how work is priced or evaluated creates a constant feeling of uncertainty for workers. The changes make it difficult for them to know if they are in compliance with the terms of service, which puts them at further risk of being blocked by the system.

These challenges have led RappiTenderos to engage in labour actions, and this raised additional concerns about Rappi's ability to surveil its workers. Participants noted that Rappi's monitoring of their geolocation is intensified when they take part in mobilizations organized by UNIDAPP, and that taking part in a peaceful protest can result in an indefinite blockage. The company also monitors them by deploying "brigadistas," people hired to carry out arbitrary verifications of RappiTendero accounts. According to the participants, the brigadistas harass RappiTenderos in various ways: they block them without justification or ask to check their cellphones to see if there is any inconsistency between their personal data and the registered account. Brigadistas also exploit the fact that many RappiTenderos in Colombia are migrants from Venezuela or other countries. They can appeal to the National Police or Colombia's

Migration officials to check the workers' identification, which results, among other things, in the suspension of their accounts (Pacheco 2020).

In view of the concerns expressed by RappiTenderos, Karisma asked participants to imagine how the platform would look if it reflected the needs of its workers. To begin with, participants said that the platform would distinguish between personal and work spaces by turning off monitoring and notifications when the application is not in use. Second, they would not be forced to keep geolocation turned on at all times, and the company would not use location information to harass them during union and advocacy activities. Third, the company would no longer punish them for choosing not to connect to work, and would respect their right to act as independent workers. They also believe that the company should establish a procedure for resolving customer complaints. Blocking should not happen without hearing the worker's account of what happened. This would resolve data and power asymmetries by equalizing the weight that Rappi puts on data produced by workers and data produced by retailers or customers. Finally, the company's algorithm should be accurate, transparent, and fair in how it calculates distances and pricing, and it should be able to properly assign the locations for jobs. As one participant says, the algorithm should work to "make their job easier."

Finally, using the technique "value versus values," the group was able to articulate their expectations and demands for Rappi. They said that workers should not feel obliged to work for a delivery company, but should "feel worthy of working" and "feel good to work in his company." This was framed as an equity concern, meaning that the company would value its workers as much as it values its business partners or customers. They also noted the importance of empathy. They pointed out that Rappi's motto is "Express with Love" but asked, "What empathy do you have with your worker? None, because you do not guarantee him anything. It's all hypocrisy in the end."

Overall, RappiTenderos understood the need to share personal data with the platform in order to make deliveries, but when they do so they want their information to be mobilized in ways that improve working conditions. As they see it, the Rappi platform sets up a clear data and power asymmetry. Workers are not treated equally to other actors within the platform, and, what is more, their information is often used in ways that negatively affect them. They noted that there is no proportionality between the three central actors within the service: customers, partners, and RappiTenderos. Information produced by delivery drivers is leveraged for logistics, marketing, and strategic planning but is not used to improve the experience of workers. Also, workers' information is treated as valid when they are subject to data capture, but it is not treated as valid when they exercise their right to complain.

Disproportionality within the system manifests through surveillance and intrusiveness at some times and disinterest and neglect at others. Interestingly, RappiTenderos do not perceive this as an issue of personal data, but rather as an effect of data-in-use. For example, while geolocation is a form of personal data, workers focus their attention not on data exploitation, but on how the notifications that result from data use invade family or intimate space. The data in this instance becomes a vector of control in the context of class relations, in that it is used to identify when workers aren't connected so that the algorithm can send them a message calling them back to work. Similarly, when the platform's interpretation of personal data sets up workers to complete an excessively long ride for little pay, workers read this as a question of ineffective management that harms their earning potential and alienates customers, rather than as an invasion of privacy.

What we learn from this is that, as an information system, Rappi sets up social relations and processes that are marked by data and power asymmetries. The system is not designed to empower workers with the information they require to make informed and profitable decisions, but instead establishes forms of control that subordinate workers. These asymmetric relations are deepened by ongoing changes to the terms of service, and by the lack of clear dispute resolution mechanisms. Taken together, this case reveals the importance of developing collective responses to algorithmically structured inequality. The Rappi workers who worked with Karisma on this process gained greater clarity about how their values and interests could be reflected in the design of an information system, and what that would mean for their daily workflow, social relations, and class relations. The next step is to mobilize as part of UNIDAPP to put these changes into practice.

REFERENCES

Arbeláez, María Angélica, Cristina Fernández, and Daniel Hernández. 2021. "Plataformas digitales y contribuciones a seguridad social. El caso de Colombia antes y después de la pandemia" ["Digital Platforms and Social Security Contributions. The Case of Colombia Before and After the Pandemic"]. *Coyuntura Económica, Fedesarrollo* 51 (December): 81–114. https://ideas.repec.org/a/col/000438/019959.html.

Athreya, Bama. 2020. "Slaves to Technology: Worker Control in the Surveillance Economy." *Anti-Trafficking Review*, no. 15 (September): 82–101. https://doi.org/10.14197/atr.201220155.

Choudary, Sangeet Paul. 2018. "The Architecture of Digital Labour Platforms: Policy Recommendations on Platform Design for Worker Well-Being." International Labour Organization. https://www.ilo.org/publications/architecture-digital-labour-platforms-policy-recommendations-platform.

Daza, Carolina, and Kruskaya Hidalgo Cordero. 2020. "'La aplicación te precariza'. Experiencias de personas repartidoras y conductoras en Ecuador" ["'The app makes you precarious.' Experiences of Delivery People and Drivers in Ecuador"]. In *Precarización laboral en plataformas digitales una lectura desde América Latina* [*Labor Precariousness in Digital Platforms: A Reading from Latin America*], 83–98. Friedrich-Ebert-Stiftung Ecuador FES-ILDIS. https://library.fes.de/pdf-files/bueros/quito/17108.pdf.

Díaz Sánchez, Gisell. 2021. "Rappitenderos de Cartagena en paro por políticas 'desfavorables'" ["Rappitenderos of Cartagena on Strike due to 'Unfavorable' Policies"]. *El Universal*. https://www.eluniversal.com.co/cartagena/video-rappitenderos-de-cartagena-en-paro-por-politicas-desfavorables-YI4082076.

Fernández, Cristina, and Juan Arbeláez. 2020. "Las plataformas digitales, la productividad y el empleo en Colombia" ["Digital Platforms, Productivity and Employment in Colombia"]. *Fedesarrollo*, July. https://www.repository.fedesarrollo.org.co/bitstream/handle/11445/3962/Repor_Julio_2020_Fern%C3%A1ndez_y_Benavides.pdf.

Mandl, Carolina. 2019. "Japan's SoftBank Invests $1 Billion in Delivery App Rappi." *Reuters*, 30 April 2019. https://www.reuters.com/article/us-softbank-investment-rappi-idUSKCN1S619Y.

Mateescu, Alexandra, and Aiha Nguyen. 2019. "Algorithmic Management in the Workplace." *Data & Society*, 6 February 2019. https://datasociety.net/library/explainer-algorithmic-management-in-the-workplace.

Máximo, Ernesto Jaramillo-Molina. 2020. "Precariedad y riesgo: Diagnóstico sobre las condiciones laborales de los repartidores de apps en México" ["Precarity and Risk: Assessment of the Labor Conditions of App Delivery Drivers in Mexico"]. In *Precarización laboral en plataformas digitales una lectura desde América Latina* [*Labor Precariousness in Digital Platforms: A Reading from Latin America*], 135–48. Friedrich-Ebert-Stiftung Ecuador FES-ILDIS. https://library.fes.de/pdf-files/bueros/quito/17108.pdf.

Negri, Sofia. 2020. "Condiciones laborales, proceso de trabajo y movilización de los trabajadores en plataformas de reparto en Argentina" ["Working Conditions, Work Process and Mobilization of Workers in Delivery Platforms in Argentina"]. In *Precarización laboral en plataformas digitales una lectura desde América Latina* [*Labor Precariousness in Digital Platforms: A Reading from Latin America*], 37–54. Friedrich-Ebert-Stiftung Ecuador FES-ILDIS. https://library.fes.de/pdf-files/bueros/quito/17108.pdf.

Pacheco, Daniel. 2020. "Rappi, el unicornio que usted subsidia" ["Rappi, the Unicorn You Subsidize"]. *El Espectador*, 18 August 2020. https://www.elespectador.com/opinion/columnistas/daniel-pacheco/rappi-el-unicornio-que-usted-subsidia-column.

Revista Semana. 2021. "Marchas hoy: Rappitenderos anunciaron movilizaciones en Bogotá y Cali este lunes 8 de febrero" ["Marches Today: Rappitenderos Announced Mobilizations in Bogota and Cali This Monday, February 8"]. *Semana*, 8 February 2021. https://www.semana.com/nacion/articulo/marchas-hoy-rappitenderos-anunciaron-movilizaciones-en-bogota-y-cali-este-lunes-8-de-febrero/202139.

Rosenblat, Alex, and Luke Stark. 2017. "Algorithmic Labor and Information Asymmetries: A Case Study of Uber's Drivers." *International Journal of Communication* 10 (27). https://doi.org/10.2139/ssrn.2686227.

Sánchez, Sheila. 2020. "Esta es la estrategia de Rappi para convertirse en la 'superapp' latinoamericana" ["This Is Rappi's Strategy to Become the Latin American 'Superapp'"]. *Forbes México*, 14 July 2020. https://www.forbes.com.mx/esta-es-la-estrategia-de-rappi-para-convertirse-en-la-superapp-latinoamericana.

Sánchez Vargas, Derly, and Oscar Maldonado Castañeda. 2020. "Obstáculos, resistencias y repertorios de movilización social de repartidores digitales en Colombia" ["Obstacles, Resistances and Repertoires of Social Mobilization of Digital Deliverers in Colombia"]. In *Precarización laboral en plataformas digitales una lectura desde América Latina* [*Labor Precariousness in Digital

Platforms: A Reading from Latin America], 99–118. Friedrich-Ebert-Stiftung Ecuador FES-ILDIS. https://library.fes.de/pdf-files/bueros/quito/17108.pdf

Vita Mesa, Laura. 2020. "Rappitenderos protestan en el norte de Bogotá para solicitar mayores garantías laborales" ["Rappitenderos Protest in Northern Bogotá to Demand Greater Labor Guarantees"]. *La República*, 15 August 2020. https://www.larepublica.co/economia/rappitenderos-protestan-en-el-norte-de-bogota-para-solicitar-mayores-garantias-laborales-3046341.

WHAT WE LEARNED FROM CRITICAL DESIGN RESEARCH

The five data literacy case studies carried out by project partners were each unique. The issues under consideration were each situated in a local context, and the workshops were convened by different organizations. Each group also used different techniques and had a different relationship with the subject matter and local participants. And yet when we look over the cases, we notice certain consistencies in how people experience data in their daily lives, and the assumptions, values, and contexts that shape how people experience data. These observations helped us begin to imagine alternative approaches to data literacy.

The five case studies evidenced that across Latin America, datafication – in conjunction with other processes such as digitization of services, platformization of business logistics, customer management through loyalty programs, and records management – is reshaping how people take part in and experience daily encounters, activities, and transactions. We see this clearly in the gradual move from neighbourhood to chain pharmacies in Paraguay, and in the fact that some participants linked the brand prestige and corporate dominance of chain pharmacies to their ability to mobilize personal data for service provision and discounts. We see this reflected in the way that young women in Peru express corporeal autonomy by relying on the internet and their friends to get information about the side effects of contraceptives and see pharmacies not as a source of health information, but as a place to transact purchases and acquire points. In Chile, we see how the ability of migrants to access a national ID card determines whether they can participate in the formal economy, marking a clear divide between citizens and migrants. In Uruguay, elderly people worry that newly introduced digital health records could medicalize aging by introducing standardized measures of health and wellbeing. And in Colombia we learned how Rappi delivery drivers are organizing against platform practices that threaten their rights as workers.

While the experience of *change* wrought by datafication is common across the cases, datafication clearly manifests in myriad ways with multiple effects. In addition, the systems and processes that people encounter in their daily lives are often designed to be invisible, such that people focus not on the data but on the convenience, efficiency, or benefits enabled by datafication. This points to a key assumption of data literacy work, which is that people need to be educated on how to manage their personal data, or how to understand socio-technical processes, when in many cases what actually resonate for them are the effects of technological change and the implications of those changes for issues that matter to them, such as family, affordability, or pleasure. This suggests a shift away from focusing on data and towards considering social relations and how they change to accommodate or respond to new technological incursions. It also calls into question a second common assumption about data literacy work, which is that the focus should be on uses of personal data, rather than narratives and norms around personal data use. Convenience is a powerful narrative and also a social norm, and data plays a significant role in how that narrative and norm function. This suggests shifting the focus from the data to the discourses and practices that surround it.

We were also struck by how conflicted people felt about datafication, but at the same time how isolated and individualized they feel during moments of data-sharing. This conflict was captured perfectly in the comment "internally I say to myself 'it's fine' even though I am questioning things," which was shared by a participant from Peru. Participants also repeatedly expressed the feeling that they had no choice but to share their personal data, because it was a condition of accessing the goods or services being provided. As a result of these dynamics, we observed that datafied systems are already well-embedded in people's daily experience, and rather than resist them, users often seek ways to accommodate them, take advantage of them, or survive despite them. This points to another common assumption in data literacy work, which is that people are coming to data in a fresh way, without the burden of pre-existing commitments, and that once they learn about data and the systems that use it, they will be able to make different choices about how they share their personal data. Changing how we manage our personal data is often well beyond what an individual can or would do, so showing them that a system takes advantage of their personal data is liable to elicit a "so what" response, or make the participant feel disempowered. It also assumes that what brings people to spaces of data literacy is a need to be informed rather than a need to construct alternatives, when in fact what many people desire are strategies they can use to take advantage of data's many new affordances.

Having said this, while some people are keen to figure out how to function within new systems (or even game the system!), many others don't fully understand the bargain they've entered into when they become part of a datafied system. A key assumption of data literacy initiatives is that the main subject of the work is the holder of personal data, and that the goal is to build up the consciousness of the individual. However, the case studies suggested that it is difficult to make sense of data bargains on one's own, especially given the fact that their effects are often social and relational rather than individual. The case studies showed that becoming aware of the full impact of the commitments we're making when we share our data is easier when ideas are shared. This suggests that groups of people need to be convened not just around particular instances of datafication such as a database, information system, or platform, but within a grounded setting that helps them make sense of why this system matters and how it shapes their social relations. It's the experience of being part of a group that will allow participants to make sense of a particular use of data, and this in turn suggests a different goal for data literacy: not the citizenship or participation of individuals, but rather generalized understanding and a basis for coordinating how data gets used.

We found that situated data literacy interventions will most likely bring together a group of strangers. There is no natural community of data holders who use the Peruvian pharmaceutical chain InkaFarma, for example, so convening this group is not straightforward. And because of this, workshop participants came to these spaces as individuals who talked about their experiences in terms of "I" or "me." However, once they came into the workshop space and began to share their experiences with each other, they began to notice their common experiences of a techno-managerial phenomenon, and through this they began to discover their "groupness." For example, in Uruguay participants came to realize that standardized health records management risks instituting a generic view of aging that medicalizes the elderly. ObservaTIC concluded that the workshops helped participants arrive at critical views on the relationship between data collection and the quality of their experience of healthcare and aging. The workshops didn't necessarily remove the functioning of algorithms, categorizations, or pricing schemes from its black box, but people became more conscious of how important those processes are to their daily lives, and came to realize how these systems organize social relations.

In total, from these observations we identified some important patterns. The findings suggested the difficulty of becoming "data literate" given the many ways that data functions in our daily lives. And this puts into question the main goals of data literacy, which are typically framed as creating a knowledge base for individual citizenship, participation, and advocacy. It also calls into question the assumption that we each need to become critically aware of how datafication shapes our lives and learn how to manage our personal data resources independently. On the other hand, it points to the potential of grounding literacy in how datafication mediates our social experience, and how that in turn shapes or changes our interactions with each other. Through sharing, the situated effects of datafication become more apparent, and it is these effects (rather than the data itself) that will influence how people respond to datafication. These findings left us with two design imperatives. First, we need language or concepts that help us talk differently about how people experience datafication. Second, we need a set of techniques for working with people and helping them see how datafication is shaping their collective experience.

Section Two

—

Data Literacy Concepts to Inform Practice

THE UTILITY OF CONCEPTS

Theory and concepts don't exist for their own sake. They offers frameworks and metaphors to make sense of the world. Those metaphors are like a flashlight that focuses our attention on certain things and leaves others in the dark. They are "designs" or "constructs" that orient our thinking and our practices. They shape what we think the problem is, and how we believe it should be solved. They orient our efforts to produce the world; however, they won't serve our purposes if they don't resonate with people or provide relevant insights.

Personal data literacy is informed by all sorts of constructs, and these orient both how people *think* about literacy and how they intervene in *practice*. For example, a common assumption orienting work on personal data literacy is that people need to learn how to be good data citizens (Carmi et al. 2020; Gezgin 2019). This is often interpreted to mean that they understand their rights and responsibilities where personal data is concerned and are able to protect themselves in the face of potential privacy or security risks. But if this isn't what people are experiencing as a result of datafication, or if citizenship doesn't offer a relevant framework for literacy outcomes, then these assumptions won't offer a productive starting point for literacy interventions. Based on what we observed from our partners' work in Latin America, we feel it's productive to consider alternative starting points for data literacy. We've summarized our views in the following paragraphs, and we offer an expanded explanation of these concepts and arguments in the rest of this section.

We argue that the information systems we belong to – what we call *Communities of Convenience* – organize our social relations in ways that are difficult to perceive or make sense of. We contribute to these processes and become part of them by sharing our personal data with them. Some authors argue that these are hegemonic processes; however, in practice, they are diverse, and we participate in them in many ways, for different reasons, and with different effects. This suggests that data literacy should focus on helping people understand how datafication shapes social structure and social relations in specific contexts. As a concept, the value of communities of convenience is that it orients data literacy interventions away from data or technologies, and towards datafication's historically and contextually specific effects on people, social relations, and communities.

It is difficult, however, to bring people together into processes that allow them to make sense of their involvement in communities of convenience. There is a complex tension between how datafied systems see us and how we think about or engage with the world around us. The experience of being multiply indexed in many communities of convenience leaves us hermeneutically and epistemologically untethered, a condition that we call the *Unhomed Data Subject.* While information systems and data policy treat the data subject like a coherent singularity, as subjective thinkers we make sense of datafication and its effects in migratory, unsettled, or ungrounded ways. Rather than assuming people occupy a unified subject position and coherent subjectivity, the concept of the unhomed data subject suggests that data literacy efforts should start from the point of unsettlement or displacement.

We need, therefore, starting points for data literacy interventions that prioritize and leverage people's diverse, opaque, and untethered experiences. Everyday acts of data-sharing are an important source of insight in critical data studies, and we also leverage them as a starting point. However, we are cautious to balance this approach with the idea of *Everyday Invisibility*. The concept of the everyday suggests that simple acts of creativity, resistance, or appropriation are just as important to shaping datafication's effects as the systems that organize data collection. As a result, some data literacy interventions aim to empower people to make different choices, embrace new behaviours, or apply creativity during everyday moments of data-sharing. This is important work;

however, we should be careful not to rely too heavily on the everyday as a source of resistance or change. Doing so assumes that datafication and its effects are already tangible and that people are able to act freely in the face of the organizing power of digital networks. In contrast, we argue that data literacy should position the everyday as a source of grounded knowledge about people's situated experiences, feelings, and opinions. This knowledge can be used to build up a mosaic of alternative knowledge bases and cultural understandings that counteract the everyday invisibility of data-driven processes.

This then brings us to the concept of *Legibility*. It is difficult to make sense of how communities of convenience organize us because their processes are black-boxed and their effects are embedded or normalized. An important goal of data literacy interventions is to make datafication's effects on our social relations legible to the people who are experiencing them. In contrast to approaches that prioritize data visualization or data understanding as a starting point for literacy, here we prioritize the legibility of data's *effects*, meaning becoming conscious of how datafication reorganizes our social relations. We argue that this work is difficult to do alone, but that by entering into dialogue with others and exploring our knowledge of everyday experiences, we begin to see how datafication challenges our values, changes our relationships, and reshapes our experiences.

In total, therefore, we argue that to achieve data literacy we need to commune, communicate, connect, make legible, and mediate, and, through these activities, enact the world in new and different ways. We produce our world through our ways of communing with one another. Becoming literate, in this view, is not the acquisition of knowledge, but rather the enactment of cultures of knowing. *Communing data literacy* is a means to counteract datafication's symbolic and material power and its ability to assert ontologizing forces. Communing does this by manifesting the epistemic and mediating power of dialogue and collective meaning-making.

In what follows, we present these five concepts, contextualize them theoretically with reference to existing concepts and traditions, and explain their relevance for thinking about and doing data literacy work. One of the benefits of this exercise is that it forces us to be transparent about the assumptions orienting our work, and offers you, the reader, a framework by which to assess the techniques that we designed on the basis of these concepts. We like this transparency because it makes dialogue and interaction about data literacy more likely and more accessible, and this excites us as academics! However, we recognize that not everyone will want to dive into the theoretical backdrop that informs our data literacy techniques. Often learning, thinking, and being happen more readily through examples, practice, and experimentation. If that's you, then we suggest skipping ahead to section 3. The explanations above give you enough of a foundation to move ahead, and you can always come back to the theory and concepts later if you find that you have further questions.

COMMUNITIES OF CONVENIENCE

We use the term "communities of convenience" to draw attention to how data-driven processes organize our social relations. When we talk about communities of convenience, we are thinking of customer databases, loyalty programs, platformized services, or the like. These are large, data-driven processes that organize people by way of digital systems, brief moments of transaction, and customer service scripts. In contrast to a community that is formed around tangible and legible material and affective relations, a digitally enabled community of convenience brings people together through intangible and invisible flows.

These data-intensive processes of assembly or organization go on whether we are conscious of them or not, meaning that in our digitally mediated world, we are part of dozens if not hundreds of communities of convenience of different kinds, often without being aware. Sometimes the assembly is being done to people, and sometimes people choose to become affiliated with these systems, but in either case the members of these communities become part of larger processes that make decisions, set prices, allocate resources, or determine who gets access. In these processes, it is very difficult to enter into dialogue with the other members, or even to know who they are. It is possible for these affiliations to provide a sense of belonging or identity, but as Gandy (2021) so clearly lays out in his work on the panoptic sort, in these instances our connections are with a database and they function according to the logics of capital. The connections that are formed emphasize efficiency and economy, and thus are the antithesis of the slow, patient, and affective work of human and ecological relation.

Because it focuses on social relations, the concept of community of convenience draws attention to the *people* who are brought together by data, rather than the data that is assembled from people. It also draws attention to how people experience social connection as a result of data power and data assemblages. This is an important point because if we lose sight of data's connection to human or ecological truths, we lose the grounding necessary for critical empirical analysis of datafication's social effects. Data is an abstraction that is generated out of and contributes to material and cultural realities, and we need to keep this insight clearly in view. When we lose sight of that, we lose sight of the "why" of critical data literacy interventions.

In this sense, the notion of communities of convenience differs significantly from the idea of data assemblages put forward by Kitchin (2014). In his work, these are defined as "a complex socio-technical system, composed of many apparatuses and elements that are thoroughly entwined, whose central concern is the production of data" (25). The idea of data assemblages prioritizes data and information products and the systems that create them rather than the people who are the subjects of datafication and objects of datafied social organization. It gets taken up in works that assess how networks are organized, or the value of the data that flows through a particular assembly of actors and technologies. If we anchored data literacy in the concept of data assemblages, we would be helping people think through their participation in a system, rather than comprehend how the system shapes their experiences and relationships.

Communities of convenience, on the other hand, focuses on the fact that people and relationships are being organized and managed in different ways depending on the intent of a data assembly. While the concept of data assembly is clearly useful and important, communities of convenience puts the focus on situated socio-technical processes, their historical effects, and our localized experiences of them. This provides the necessary grounding to think through how data power and data assemblages become operative in our daily lives, how they organize our social relations, how we feel about that, and how we might organize ourselves in response to these processes. It therefore offers a very different starting point for thinking about data literacy interventions.

The idea that people are actively assembled and influenced by social or technological processes is not new, nor is the notion that we are often unaware of these organizational processes. However, the role of datafication in these processes has fundamentally changed how we experience them. Here we can draw useful insights from critical advertising studies. Twenty-five years ago, Miller and Rose argued that "[m]aking up the consumer entailed simultaneously making up the commodity and assembling the little rituals of everyday life which would give that commodity meaning and value" (1997, 17). This involved careful psychological work that formed the basis of "human technologies" such as tasting profiles or seasonal trends that could be used to mobilize consumers to make connections between their passions or proclivities and the features of the goods they might buy. The consumer was positioned as a lone subject who, faced with an overwhelming set of choices available in the marketplace, could be oriented according to a "cartography of consumption" (30). Clusters of offers might emerge around affinities, and these were designed to help consumers select between, say, opera and show tunes, in ways that both drew on and produced a sense of belonging and connection. In this model, people were positioned as autonomous consuming subjects; they were being influenced, but the choice of what type of music to listen to, if any at all, was ultimately their own.

What differs now is how we experience and think about the relationship between efforts to organize people or construct identities and people's ability to comprehend and exceed these forces. In digitally mediated market spaces, consumption is increasingly organized as a way to manage consumer outcomes (Zwick and Bradshaw 2018). Organization is carried out according to inferences gathered through analysis of data, sometimes by the algorithmic systems that these inferences inform. These processes happen behind the scenes, so while our everyday experience is one of convenience and flow, in the background digital decisions determine what products we see, what scripts we read, or what instructions we receive. In total, we are seeing a shift from the sovereign client free to make choices in the presence of subjugating pressures, to the reflexive user whose freedom of choice is an instrument of their subjugation (Bęckett and Nayak 2008) and also a means of their exploitation (Zwick, Bonsu, and Darmody 2008).

Fourcade and Healy argue that these systems work to reproduce social class by leveraging insights about our *habitus*. As they say, "when markets see, this matching feels all the more natural because it comes from within – from cues about ourselves that we volunteered, or erratically left behind, or that were extracted from us in various parts of the digital infrastructure" (2017, 17). As we go about our daily business, we willingly, unwillingly, and/or unknowingly share insights that inform the creation of these categorizations, and they in turn organize our social relations. This leaves real people in the real world to figure out the new "markers of access to resources, services and power" (Haggerty and Ericson 2000, 613), which, given the nature of digital systems and increasingly artificial intelligence, can be difficult if not impossible to do.

In this sense, a community of convenience is more than just a digital interface or membership in a program. The organizers of these relations use their observations about us to set up a process or deliver a good or service in ways that will allow them to shape our social experiences. These processes are socio-technical in nature, meaning that they involve codification at the level of both computational systems and social processes. A loyalty program, for example, is made up of a database of members, a marketing program, staff trained to follow client-relations scripts, and members who are taught and incentivized to use their membership card when they shop at certain locations. Affiliation can take many different forms, including compulsory, voluntary, exclusionary, or conditional, but regardless, the members of data assemblies do not get to see the membership roster, and typically won't be conscious of other members unless there is a mechanism to connect with "friends," or someone makes an independent effort to convene or locate other members. Either way, members typically have little say over the terms of their involvement, even as they are unconsciously contributing to the terms of their participation and its local effects.

Given that we face multiple digital pressures for assembly, we may sign up for information systems as autonomous subjects, but we populate them as transient bodies. For Deleuze (1992) digital systems make it so that people are no longer productively understood as indivisible units with an autonomous will who can form into masses to project their desires. Rather, they are better seen as fractal bodies enmeshed in digital networks, from which epistemic resources can be abstracted and organized, and onto which the expectations of those abstractions and forms of organization can be projected. Deleuze calls these bodies "dividuals" to mark a distinction with earlier ways of understanding the basis of social organization. As Martinez (2011) explains, the body which was once "targeted by disciplinary techniques once it entered a panoptic enclosure is now abstracted as digital codes," and this enables "many forms of breaking down and representing the individual." The resulting dividual can then "travel across the overlapping networks continuously and instantly" (208). This is akin to saying that where workers were once indivisible and permanently assigned to specific tasks, today they have been flexibilized such that they can be organized according to the shifting will of digital systems. Haggerty and Ericson (2000) further explain that by "abstracting human bodies from their territorial settings and separating them into a series of discrete flows," people can be "reassembled into distinct 'data doubles' which can be scrutinized and targeted for intervention" (606). For example, consumers can be sources of multiple potential streams of revenue at the same time because each of their fractal selves can be targeted differently.

This makes us susceptible to the logics driving all sorts of different projects of data-driven organization. This type of experience is captured by Martín-Barbero, who explored how, in a digitally networked era, "the body itself is inserted into life worlds and into the instability of senses that configure an individuality that suffers from constant identitary fragility" (2011, 116). That is to say, we can't know ourselves as a particular identity, because we must have many different identities depending on what a system needs from us. Because our cultural reality is being mediated by many different forms of affiliation, we develop the individualized coping mechanism of being accommodating to the various different forms of being that are required of us. Flexibility and adaptability become useful coping mechanisms for working through these demands (Smith 2012). They give us the greatest chance of developing a personal sense of self in the context of many different systemic demands, even as it exposes us to the risk of exploitation. It must also be noted that this same flexibility allows people to be differentiated within and between themselves and across time in how they navigate the many different expectations of their various affiliations. This double-edged sword at the interface between socialization and enactment raises important questions about the nature of the contemporary data subject, which we explore in greater detail in the section about the Unhomed Data Subject below.

We value the idea of communities of convenience because it grounds empirical analysis of datafication's social effects, and their implications for the exercise of power in contrast with totalizing narratives. For example, some twenty years ago Haggerty and Ericson argued that pervasive digital systems had begun to coalesce into a surveillant assemblage that "operates by abstracting human bodies from their territorial settings and separating them into a series of discrete flows" which can then be "scrutinized and targeted for intervention" (2000, 606). This is mirrored in Gandy's notion of a "panoptic sort" that processes people's information to "coordinate and control their access to the goods and services that define life in the modern capitalist economy" (2021, 21). Similarly, Couldry and Mejias argue that data colonialism ensures "compliance within interlocking systems of extraction in everyday life. These systems are so many and, taken together, so encompassing that they risk governing human beings in just as absolute a way as historical colonialism did" (2019, 6).

However, to say that these changes manifest in cohesive and uniform ways is problematic. When people are told that they are facing powerful hegemonic forces, they often feel profoundly disempowered and lose hope. Meanwhile, as was pointed out in our review of cross-cutting themes from the case studies, people's localized experiences of datafication are specific and varied, and their reasons for linking themselves to socio-technical systems are complex. What is more, many of the data-intensive systems that we experience in our daily lives work hard to keep data carefully siloed. Indeed, in the United States, efforts by like-minded people to mobilize and combine different *open* data sets in support of disaster mitigation (Young et al. 2020) or environmental stewardship (E. Williams 2023) have run into all manner of obstacles (Mattioli 2017). It also stands to reason that in many digital industries, data sets constitute a core competitive advantage and thus organizations are loath to share them. Uber has famously fought off efforts to compel greater access to the fantastic data resource produced through the activities of its drivers (McBride 2014). This does not even begin to get into the privacy and security laws, consent mechanisms, and global data management norms that compel organizations to protect their databases. All of these examples tell us that hegemonic forces express themselves in diverse, nuanced, and situated ways.

So, in total, it makes sense to have concepts that orient understanding of how specific instances of datafied assembly transform social experiences in a particular context. Totalizing narratives leave us with little space to make sense of or respond to everyday experiences of these changes. For the purposes of developing data literacy interventions, we need techniques that explore social dynamics, and that leave room for people to articulate and contemplate their immediate observations, experiences, or feelings. The notion that communities of convenience use our data in a variety of ways to organize us opens the door to a grounded analysis of datafication's everyday effects, and provides a more relatable starting point for developing grounded data literacies.

UNHOMED DATA SUBJECTS

How should we think about and engage with people who are suspended within digital communities of convenience? In our view they are subjects who lack clear grounds from which to make sense of changes they're experiencing in their social relations as a result of datafication. The term we use to describe this is "the unhomed data subject." Here we are mobilizing "unhomed" as a metaphor to poetically express the experience of having one's subjectivity mediated by multiple communities of convenience.

We feel it's important to have a clear vision of how subjectivity functions under datafication for two reasons. First, a person whose data is collected by digital systems is known as a "data subject." For instance, in the EU's General Data Protection Regulation (GDPR), the data subject is an individual who can be singled out or identified by way of data they've shared. The issue with this definition, however, is that it understands a person in terms of what a computer or analyst can know about them. As the Privacy Settings Collective explains, "The data subject is, in other words, the amalgamation of the analyzable metrics, statistics, and crude figures that we become when registered and processed by legally permissible regimes of data rendition" (2023, para. 3). This over-attention to the sense-making capabilities of the machine leaves us without a way of understanding the human experience of sense-making in a context that is mediated by machines. Second, literacy concerns how we "read" and make sense of the world around us (Gee 2020), and our experiences of subjectivity are changing in fundamental ways under datafication that challenge our existing theories of the self (Braidotti 2019). We therefore want to introduce concepts that help us understand the human experience of sense-making in a context that is mediated by machines, and that draw attention to how people inhabit ways of knowing or develop self-awareness within a data-mediated sensorium.

Before we do that, its helpful to back up and review the concept of subjectivity. It refers to what we know, feel, or perceive from the perspective of our sentience, and therefore it is linked to our consciousness, self-awareness, and personhood. By "subjectivity," we mean people's cognitive and emotional positionality within our mediated sensorium (building on Martín-Barbero and Rincón 2019), and by "subjectivization," we mean the processes of consciousness-construction that happen within such a space. We therefore understand a sensorium to be a socially constructed and politically stratified regime of perception, and this means that subjectivity is mediated and experienced in uneven ways. As a concept, subjectivity orients our thinking toward how people make sense of their experiences, and what that means for their behaviour, desires, values, and actions. It also encompasses how different ways of knowing are prioritized, valued, or promoted in a given regime of perception, and how others are devalued or marginalized.

In the contemporary datafied sensorium, it is not so much the fact of data that matters to how we experience the world. Clearly, data and datafication have been around for a long time (Rosenberg 2013). Rather, what matters is how datafication is organized to interpellate our subjectivity. The duality between how we experience datafied systems and how we think about or engage with the world around us sets up a complex relationship where subjectivity is concerned. In philosophy, we are subjects in the sense of observing the world from the point of view of our biases; however, in data studies, we are the subjects of the systems that observe us. In the former case "subject" refers to the core of our being, our thinking and feeling essence. In the latter, to be a "subject" suggests that we are regulated by, or "subject to," a system that tries to know us. As Couldry and Hepp (2018) put it, in a mediated world, "[b]eing 'someone' shifts from being associated with a certain quality the self and others can abstract from the stream of everyday habitual action, to being a continuously managed 'project', that is, an 'external' responsibility of the self towards the social world" (146).

Our subjectivity – how we sense or perceive the world – emerges in relation to the world around us; however, our thought processes are also mediated, and this has implications for our being, knowing, and doing. Therefore, how we know and how we experience self-knowledge are tied up with how we are known.

As a concept, the unhomed data subject tries to get at this relationship. Data's mediating forces shape our experiences in ways that shape our positionality – and ultimately condition our subjectivity, which is to say, our sense of self, expression of agency, and ability to project futures. Communities of convenience use data to structure our social connections in particular ways at the same time as they curtail our capacity to understand how we relate to each other. Couldry (2023) calls these processes anti-hermeneutic, which is to say, forces that impede interpretation and culture-making. As he says: "What is engagement? It sounds as if it might be exactly the space of popular interpretation and creativity … But it is not that at all. It is an instrumentalized version of everyday emotion and connection that is most likely to generate moments of attention" (664). Thus, we see the unhomed data subject as a digitally connected but hermeneutically displaced person who finds it difficult to make sense of their experiences as a result of how the epistemic field has been organized (Fricker 2007).

We turn to Bhabha and Dubet to further develop a theory of the unhomed data subject because they each offer concepts that illuminate the relationship between mediation and the knowing subject. Postcolonial literary scholar Homi Bhabha (1992) mobilized the idea of unhomeliness to highlight the dominance of colonial thinking in literary theory, and to capture how it marginalized both the experiences and the narratives of subaltern peoples. In Bhabha's view, "the great homes of English Literature" (142) transmitted cultural domination and, in doing so, created epistemic or psychological refugees among those whose experiences or ways of being fell outside of the images projected by this powerful literary narrative. In building on Bhabha's work, we draw an analogy between the literary fictions that informed his thesis and the socio-technical systems that mediate the contemporary sensorium. What is to Bhabha the house of fiction is to critical data studies the techno-aesthetics of social organization. To put it another way, a colonial apparatus can be deployed in literature as it is in databases. In either case, we can think of these forces as reproducing dominant ways of knowing, and as creating an unsettled, migratory, or ungrounded experience of being.

This unhomeliness of living with and through data results not just from the fact that people's real experiences and grounded understanding exceed dominant ways of knowing, but also from the fact that this dominant form organizes our subjective experience in particular ways. In this sense, Dubet (1996) argues that as unitary roles have been displaced by diverse experiences, our subjectivity has come to emerge in and through the administration of the various logics we encounter in the many fractured expressions of modern life. That is to say, our sense of self emerges through moments of meaning-making as we navigate and negotiate the many systems that we encounter on a daily basis. From this point of view, if datafied systems are multiple, atomizing, isolating, and difficult to make sense of, then subjectivity emerges through how people "make themselves up" (Kear 2017) as they navigate tensions between the virtual forces of world-making and their actual experiences. This means that in a datafied sensorium our experiences or ways of being will be migratory, unsettled, or ungrounded, and the streams of information we give off will be multiply situated, in stark comparison to the apparent unity of the data subject that is captured by any given information system.

Rather than seeing displacement as a liability, Bhabha centred unhomeliness as an alternative approach to enunciating history. In this view, the present is "a form of temporality that is open to disjunction and discontinuity and sees the process of history engaged, rather like art, in a negotiation of the framing and naming of social reality" (1992, 144). Thus, he suggested relating the "traumatic ambivalences of a personal, psychic history to the wider disjunctions of political existence" (144). He further explains that this entails "a shift of attention from the political as a theory to politics as the activity of everyday life" in which "banalities are enacted … But it is precisely in these banalities that the unhomely stirs" (149). Similarly, we can say that unhomed data subjects "dwell" in the disjunctures and displacements created by datafication, and in doing so they produce an alternative aesthetic which is not a transcendent or ontologizing truth, but rather an imminent and epistemic experience of daily datafied realities that are constitutive of "a" truth or "their" truth.

In other words, we dwell in the everyday experience of navigating the multiple exigencies of datafied systems. This can be seen as a challenge as well as an opportunity. It is a challenge insofar as information systems displace or occupy our potential for self-knowledge by distracting us, making social organization opaque, and isolating us from each other. But it can be an opportunity, drawing on Braidotti, insofar as it creates possibilities to materialize posthuman experiences. As she explains, this type of subjectivity is "relational and hence in constant negotiation with multiple others and immersed in the conditions that it is trying to understand and modify, if not overturn. A philosophy of immanence, or of situated perspectives, implies an epistemological obligation to reach adequate understandings of the conditions of one's existence" (2019, 41). From this point of view, when subjectivity is released from the obligation to centre human existence and instead becomes the goal of human interaction, then being unhomed can become a productive and transformative space rather than a source of either egotism or subjugation.

Having said all this, it is important to pay attention to some important caveats. First, Bhabha's detractors argued that his epistemic take on unhomeliness paid insufficient attention to the material aspects of the colonial experience. For example, Benita Parry (1994) took Bhabha to task for replacing dispossession with epistemic violence, and for focusing on the semiotic terrain instead of material violence. When we focus too much on our experience, we risk overlooking the material conditions that produce it or result from it. Interestingly, critical data studies marries the forces of dispossession and epistemic violence to each other, arguing that it is precisely by dispossessing people of their personal data that data assemblies assert epistemic forms of coercion and control (Ricaurte 2019; Thatcher, O'Sullivan, and Mahmoudi 2016). Our experience of epistemic violence is precisely an experience of material violence. Therefore, we see the unhomed data subject as being in need of data literacy interventions precisely because socio-technical or techno-managerial processes of assembly are illegible and unknowable, and because people lack spaces from which to make sense of or respond to them.

Second, it is important to recognize that not all data subjects are in an equal state of unhomeliness. People's experiences of datafication are not evenly distributed across groups. In line with previous literature, these processes of subjectification ought to be understood as an expansion and replication of existing logics of power (Milan and Treré 2019; Ricaurte 2019; Sastre Domínguez and Gordo López 2019; Segura and Waisbord 2019). Communities experience the mediation of the sensorium differently according to their experiences of gender, ethnicity, or class – categories that are reinforced and expanded by datafied systems as data is mobilized to read the world into being (Scott 1998). Meanwhile different communities will engage in meaning-making from the point of view of their historical, social, and cultural experiences. This unequal burden of subjectification and world-making highlights how engagements with datafication are not the same for everybody.

In total, the notion of an unhomed data subject calls attention to people's multiple, serial, and ongoing efforts to author and constitute domains of being across the multiple platforms, algorithms, databases, archives, and systems we experience as we move through contemporary life. What this means for data literacy is that we can dwell in our everyday experience of navigating communities of convenience. Narratives about people's experiences of data-driven affiliation need not start from the assumption of coherent and autonomous subjects, nor from the assumption that we experience freedom and choice, nor even from the idea that the workings of technology must be understood by the people who are subject to it. We can start the story from the place of flexibilization, displacement, alienation, or disaffection, and from there think about how this organizes our daily encounters, and what this means for how social relations could be.

The notion of an unhomed data subject provides a grounds from which to enunciate or narrate the experience of being multiply indexed into various communities of convenience. It gives us the grounds from which to understand that communities of convenience use our data to organize us into value propositions that may be inequitable or undesirable. They can do this because these structures position unhomed data subjects as an anonymous and fractured user base. Thus, in contrast to Bhahba's geopolitical displacement, in the contemporary moment our internalized feelings of foreignness take on a geo-technical guise. These material arrangements integrate us into socio-technical knowledge infrastructures that mediate both the regime of perception and our cultural experience. And yet, even though this is an epistemic project, and even though we are essential to producing and performing the knowledge-culture that emerges from it, we face barriers to understanding it. In organizing socio-technical systems on the basis of corporate agendas, communities of convenience determine the values that will inform our experience of community, and they in turn create social expectations that cause us to reproduce the very anonymous and fractured user base that allow communities of convenience to function. Thus, in total, our experience of being unhomed is directly related to our experience of being organized into communities of convenience.

EVERYDAY INVISIBILITY

Everyday experiences of datafication have been widely explored in critical data studies literature (Burgess et al. 2022; Gebre 2022; Alaimo and Kallinikos 2017; Pink, Lanzeni, and Horst 2018; Kennedy 2018). Everyday practices are targeted for data collection by actors engaged in data capture and surveillance, so several authors have addressed datafication by asking people how they feel about or experience data-gathering. While we appreciate and value these works, in this section we want to call attention to a common assumption about the everyday, which is that datafication and its effects are tangible to us, and that we respond to these effects as autonomous and conscious individuals who have clarity about how to make sense of things. When we shift our attention to the invisibility of processes that we take part in every day, the opacity of the bargains that people enter into, or how these systems change our social relations, then we arrive at a different starting place for data literacy interventions.

It's one thing to say that technologies gather our data in everyday life, which is true. Our social, cultural, and economic processes have become so widely digitally mediated that we are leaving trails of data behind us with practically everything we do. It's quite another thing to talk about data and "the everyday." The latter is a theoretical response to the idea that culture emanates from taste-makers and is transmitted through carefully curated expressions of art, etiquette, or style; or that culture is dictated by those who occupy positions of influence due to structural racism or patriarchy. Theories of the everyday argue that culture is constructed, transmitted, and transformed in the social and material practices that make up our everyday lives, and that as a result, culture is a site of struggle and a potential space of resistance to structural inequalities (R. Williams 1961). From this perspective, data can be studied not just as a tool of capital or governance, but also as something that comes into being through our everyday practices. As Burgess et al. argue, "data is cultural – that is, data is the result of, and an agent in, processes of meaning-making and struggles over values" (2022, 36).

Everyday practices have been theorized to open up space for people to engage in strategies and tactics of cultural creativity and resistance (de Certeau 1984) and to domesticate the information flows and cultural influences of media (Silverstone 1994; Martín-Barbero 1993) through processes of cultural or technological appropriation (S. Morales 2009). As Pangrazio and Selwyn explain, actors in positions of power give us access to strategies that we can use to take better advantage of datafication, such as managing permissions or leveraging access-to-information laws (2023, 148). We can also mobilize creativity and innovation as a tactic of resistance or appropriation through acts of data disobedience (Brunton and Nissenbaum 2016) or artistic responses (D'Ignazio 2017). Even though tactics are always mobilized within an already-established social field and in the context of power structures, these moments of free experimentation give users a way to resist imposed systems of meaning (de Certeau, Giard, and Mayol 1998, 264).

When people engage with the everyday through strategies or tactics, they are said to "domesticate" the forces of cultural domination and appropriate media content to serve their own purposes. As Silverstone explains, "The world of everyday life is a world of consumption certainly, but consumption has to be understood as productive. Buying, using, reading – none of these activities leaves the subject, the object or even the system untouched" (1989, 79). Similarly, Martín-Barbero's famous epistemological shiver grapples with the question of how different audiences come to interpret the same content in wildly different ways. As he explained: "Consumption is not just the reproduction of forces. It is the production of meanings and a site of struggle that does not end with the possession of the object but extends to its uses" (1993, 214).

The trouble with the everyday for our purposes is that it tends to overstate the extent to which people know, understand, influence, or resist datafied processes, or, more importantly, the forms of social relation that result from datafication. For one thing, given how pervasive data is, we are constantly navigating and negotiating the various systems, processes, and platforms that attempt to assemble us into the spaces that manage our choices, desires, purchases, and movements. Meanwhile, each datafied effort to organize people comes with its own systems, scripts, and conducts, costs and benefits, attractions and pain points. At each turn we are asked to make sense of efforts to organize us and we must make decisions about how to respond. We are conscious of some of these processes, such as the loyalty programs that we decide to sign up for, and we may choose to be a part of them due to their convenience or perceived benefits. Other processes are obligatory, such as government-mandated health records databases, and yet others create spaces of exclusivity by setting conditions on participation. But regardless, in most cases these systems are deliberately designed to blend seamlessly into our everyday experiences so that they become unnoticeable, and their processes are typically black-boxed so that we can't know the logics that inform their data use.

This makes it difficult to assume that people navigate the many logics that organize us as autonomous decision-makers who have the time, means, and liberty to creatively resist the logics that they're encountering. Making sense of the cultural implications, value propositions, terms of service, or user agreements we confront when engaging with a community of convenience is demanding work. Impenetrable legal documents or random popup screens make a mockery of the notion of an informed consent process. People may lack the means necessary to make sense of what they are agreeing to. And, even if people are totally clear on what is being asked of them, they often feel compelled to agree simply because they need or want access to the service being provided. In total, the idea of everyday data simplifies what is actually a complex state of affairs by treating individuals as if they were coherent, autonomous decision-makers imbued with a strong sense of self and the ability to express themselves creatively within a context that is intangible and unknowable.

Second, as members of datafied communities of convenience, we simultaneously produce our world through the act of consuming it. We consume the cultural expression of datafication *and* the technologies that deliver it in very solitary ways while staring at screens, cocooned by our earbuds, or during interactions that are momentary and transactional. As we consume that content or engage in these transactions, we are producing the data that is used to feed us yet more content or inform yet more transactions. When we express ourselves in a digitally mediated system we produce data that organizes us according to categories that are beyond our reach. As a result, we unknowingly contribute to producing the controls that organize our affiliations, and we do so according to logics that we can't see or know. Thus, "[i]f during centuries technology was considered a mere instrument, today it is already well on its way to becoming reason [itself]" (Martín-Barbero 2011, 110). As Pangrazio and Selwyn point out, "striving for better forms of data-driven technologies throughout our everyday lives actually provides support for the generation and processing of more data" (2023, 157), and this means that even if people were tactical in their everyday acts, they would still be driving forward invisible processes that are difficult to know.

Finally, since these systems are embedded, intangible, and unknowable, it is hard to sustain the idea that we can easily "domesticate" how datafication shapes our social relations. Consider Katherine's experience of media consumption in her childhood home. When she was young, Katherine's family had a large television set in the family room. But when her mother re-joined the workforce in 1980 after having three children, the family purchased a 13-inch television for the kitchen. It featured two channels and sat perched on top of the fridge. When Mom got home from work she always watched *Three's Company* at 5pm followed by *Wheel of Fortune* at 5:30 while she made dinner and leafed through the local newspaper (*The Brandon Sun*). Then at 6pm the family – Mom, Dad, and three kids – would gather at the kitchen table to eat dinner while watching the CBC evening news. The news of the day – national and personal – was always discussed and made sense of in this family setting. In this sense, the family was able to 'domesticate' the news, and this practice contributed to producing a cultural expression of family.

It's not hard to extend the idea of technological domestication to Katherine's two nieces and two nephews aged 11, 13, 18, and 22 who have grown up in the age of social media. Her brother and sister-in-law's house doesn't have a television. Instead, each family member engages with content on their phone or iPad, sometimes commenting about what they're watching with each other, either in person or via Facebook or WhatsApp. This suits a family dynamic in which six people each have different school/activity/work/social schedules, and the parents, who work primarily from home, structure their schedules around a range of commitments. In this example, experiences and outcomes of meaning-making are different, and patterns of subjectivization are different, but the overall idea that mediation and related processes such as appropriation and adaptation are suspended in social relations still holds true.

It's more difficult, however, to extend the theory of technological domestication to communities of convenience. The members of these apparatuses are unknown to each other and these systems are designed in ways that would make it difficult for members to organize collectively. It's not that communities of convenenience don't exist as a social relation – they of course do – it's that these socio-technical systems are organized in ways that make them intangible and, to a great extent, unknowable. "To domesticate" implies agency, but can we really say that people domesticate systems if those systems are intangible and unknowable (Bridle 2018)? In addition, as noted above, we pass through many different spaces of datafication that each organize us differently, and we experience many different demands on the streams of value we produce, so it is difficult to track and engage with these processes coherently. This suggests that there are few clear grounds from which to "domesticate" personal data collection, or "enact" algorithms (Siles 2023) or associated technological processes such as platformization or loyalty programs.

There's no denying that we experience datafication as everyday moments of technological and commercial interaction, and that these moments offer us a "way in" to understanding how datafication reshapes our lives. However, we need to be careful about how we situate "the everyday" in data literacy interventions. In critical data literacies work, the everyday is sometimes situated as the goal of interventions (for example in Pangrazio and Selwyn 2023). The idea here is that once we understand the concept of datafication, why this is done, and the possible implications, we can engage the everyday as a space of cultural creativity and through this exercise our new critical insights. This situates the everyday as a space to enact the outcomes of digital literacy.

But we as researchers would rather take up the everyday as a space of grounded, participatory inquiry (Reilly and Flores forthcoming). When the processes, technologies, and criteria that gather data and organize our social relations are embedded and invisible, then the everyday offers a productive starting point for the production of spaces to know differently. Datafication and its effects on social relations may be difficult to make sense of. But what people know for sure is what they experience in their everyday lives. We know how we feel and what we think about our experiences, and those thoughts and feelings offer a grounded truth. As an added benefit, the everyday grounds our understanding of data in local experiences and cultures, thus avoiding the problem of assuming what the parties to data literacy interventions know about data. By focusing on our everyday experiences, feelings, and thoughts, we stop working with data and systems that were generated by actors in positions of power, and instead generate data that can create the ground from which to produce the world.

LEGIBILITY

Because its processes are seamless and black-boxed, datafication is very difficult to see, and because its effects are embedded and normalized, the changes that datafication creates in our personal lives and social relations are difficult to make sense of. This means that data literacy interventions should focus on making the effects of datafication legible, which is to say, clear enough to read. This approach differs from what is often recommended for data literacy interventions. Often activities start with data visualization (Markham 2020; D'Ignazio and Bhargava 2018) or data understanding (Pangrazio and Selwyn 2023), and then move on to reflection and critique. We argue that the main goal of literacy interventions should be to "read" and become conscious of the social arrangements we are suspended in as a result of datafication, and for this to happen those social relations must first be made legible. We also argue that this is difficult to do alone. By entering into dialogue with others, we can make changes to our relationships and ways of being more legible, and from there become data literate.

Promoting data literacy is not so simple as teaching people the "data alphabet." Doing so would not make them magically able to "read" technology or "read" socio-technical systems, and it certainly wouldn't help them understand why and how the uptake of data in organizational and decision-making processes matters to subjective and social experiences. Rather than start from data, we need to start from its effects and how we experience them.

This is not a straightforward proposition since datafied systems are themselves networked processes of meaning-making. They are dynamic, and particularly with the introduction of artificial intelligence, this means that their meanings and effects move and change. Meanwhile, people's experiences of datafication and its effects are many and varied, so achieving legibility requires contemplation of many different perspectives. As Bridle argues, "To deal with the network is to deal with a Borgesian infinite library and all the inherent contradictions contained within it: a library that will not converge and continually refuses to cohere. Our categories, summaries and authorities are no longer merely insufficient; they are literally incoherent" (2018, 249). As a result, data literacy is less the acquisition of skills or critical capacity, and more learning to engage in social processes of legibility-making. This is especially so to the extent that our experience of datafication contributes to isolation and disaffection. In total, legibility isn't a cognitive problem, but rather a social one. We need to first examine how and where we stand in relation to each other in the context of datafied systems. In this sense, literacy is not just something people do in the world and society (Gee 2020) or with their communities (Street 2003), but the manifestation of a new aesthetics of mediation and appropriation (Cepeda 2017; Martín-Barbero 1993).

Legibility-making should focus on inward explorations of our engagements with datafied systems in order to better understand (and even transform) outward connections to these structures. That is, we ought to explore the real experience of abstractions that arise as we engage with data rather than the production of abstractions. Drawing on Braidotti, this requires dialogue among people who do not know that they are subject to similar forces. Referring to the political economy of knowledge production, she says, "the crucial problem is the different speeds of de-/re-territorialization by bio-cognitive capitalism and the toxic saturation of the present it enacts, to the detriment of the actualization of the virtual. The violent erasure, or passive-aggressive blockage, of our collective desire to express and materialize virtual potentials affects both subject formation and knowledge practices in society. They actually exhaust us" (2019, 89). What we take from this passage is that multiple and continuous processes of affiliation, given their various efforts to organize us and extract value, leave us unhomed as well as disoriented. Because we are so busy grappling with serial processes of assembly, it is difficult to make sense of our daily experiences. Our minds are kept busy working out how to create spaces of comfort in which to nurture an unencumbered sense of self. As a result, we lack the time and capacity to make our experiences legible, let alone to imagine the world differently from how it is.

This raises the question of how people become individually or collectively conscious about their datafied lives, since without consciousness there can be no legibility, but without legibility there can be no consciousness either. Brazilian scholar Paulo Freire first introduced the idea of *conscientização*, or "conscientization" in English (Macedo 2022), as a way to raise awareness about our social and political realities (Freire 1970) by identifying "the conditions or situations that set limits to life, making it less human than it could viably be" (Ramalho 2022, 277). But as noted by Macedo (2022), the goal of conscientization is also to provide critical thinking tools that prevent the internalization of structures of oppression. To achieve this, Freire argues for "a critical comprehension of man [*sic*] as a being who exists in and with the world … It is as conscious beings that men are not only in the world, but with the world, together with other men" (1970, 453).

Consciousness and legibility clearly go hand in hand; however, we cannot effortlessly expand the Freirian concept of conscientization to our contemporary experience of communities of convenience. Freire's work assumed a cohesive subject of oppression within fixed hegemonies and thus he felt it was possible for people to achieve an objective reading of their surroundings and acquire an enduring form of empowerment. Regarding consciousness he wrote, "It is only by the way of an understanding of the dialectical unity, in which we find solidarity between subjectivity and objectivity, that we can get away from the subjectivist error as well as the mechanical error. And then we must take into account the role of consciousness or of the conscious being in the transformation of reality" (1985, 154). As unhomed data subjects the members of communities of convenience lack a clear ground from which to develop "solidarity between subjectivity and objectivity," so we need additional frameworks to inform our vision of legibility-making.

Turning to bell hooks (2014), conscientization emerges through the ongoing formation of relations of solidarity, which is helpful. A consciousness-raising group "encourages widespread sharing of incidences that might otherwise be considered minor within the larger feminist movement. Aside from airing grievances, this process also serves as a point of empathy and community building" (Brevins 2018, 93). In her work, hooks showed how consciousness-building within feminist collectives revealed the intersectional nature of patriarchal oppression and informed its ongoing analysis (hooks 2014; 1986).

However, as Law (2004) points out, as we engage with the multiple instances in which data assembles us, multiple realities are turned into place. Unlike feminists who came together around their experience of patriarchy, our data subjectivities are not the sum of our experiences of socio-technical systems, but rather a complex mesh of the multiple data doubles they proliferate plus the multiple other power structures at play. As we navigate these realities, we are left to figure out how to construct worlds for our subjective experience even as we allow data assemblages to extract streams of value from our data doubles.

So how can we develop consciousness around our experience of datafication under these conditions? Or, as Bridle (2018) asks, how can we properly describe our experience of data if we cannot see it clearly?

Here we find it helpful to revisit the work of pragmatist philosopher Charles Sanders Peirce (1955), who offered a theory of how we come to believe that something is self-evidently true. Here consciousness, both individual and collective, is operationalized via three interconnected divisions: unmediated and immediate ideas, experiences, or feelings (which he called "firstness"); shared representations and conventions that capture what is known ("thirdness"), and the reactions or relations that happen in actuality as we make sense of our experiences ("secondness"). According to this view, to reach a reading of one's experience (that is, to move from experiencing to generalizing), we require external elements that enable sense-making (Houser 1983). Extending the work of Peirce, in order for unhomed data subjects to visualize how datafication's indexing and affiliations structure their day-to-day experiences, they need opportunities to become estranged from the representations and conventions that capture what is known. That is, we need to open up spaces to get in touch with and express the firstness of our ideas, experiences, and feelings, and to explore and enable the secondness of our reactions and relations.

Drawing on the Latin American tradition of EduCommunication, we propose dialogue as a means to ground these processes of consciousness building about how we collectively inhabit information systems.

EduCommunciation argues that we come to understand concepts through explaining them to others, and it is in making something truly apprehensible to others that it becomes clear to ourselves (Coslado 2012, 165). Kaplún further argues that "it is through dialogue that counter-arguments, gaps, weaknesses and contradictions appear in ideas and notions that previously appeared coherent and solid; and in this way one's own thoughts begin to form, which would be unlikely without interlocutors, either present or distant" (1998, 214). Building on these ideas, it is easier to see our opaque entanglement with datafied systems if we get together with others and compare notes. This involves crafting exchanges across unnatural social alliances that materialize relations which would not otherwise be apparent. When we collectively point our lights at specific points we can illuminate larger realities. Through this process of building consciousness, our entanglements with datafied systems are made legible to us. Materializing the real-world relationships between people (i.e. putting them in a room together) is a first step in making invisible social structures legible.

It is unlikely that people in a conversation will be unanimous in their experience of communities of convenience, or indeed their assessments of them. Processes of inclusion and exclusion that are already present in social contexts get magnified through data-driven systems (Segura and Waisbord 2019; Milan and Treré 2019; Ricaurte 2019), and the types of data doubles that get created, and their implications for our lived experience, depend on how this power is expressed. Some people are particularly affected by the logics of information systems due to their nationalities, gender, or class, to name a few – categories that are reinforced and expanded as systems seek to organize our affiliations in ways that serve their interests. This drive targets different groups unevenly, sometimes by excluding people altogether, and at other times by including them in ways that categorize them for the sake of exploitation. The unequal burden of world-making highlights how engagements with datafication are not the same for everybody.

Nonetheless, it is important to sustain conversations that actualize knowledge and empathy because in realizing this act, participants contradict the instrumentalizing and extractivist logics that so often orient communities of convenience. These conversations may give rise to proposals for specific projects from all or some participants, or they may not, but this does not actually matter, because either way, people will walk away with a better sense of how to situate themselves on the terrain of unhomely forces, and in the process they will have performed the world differently than intended by the forces of assembly. As Braidotti has said: "Recognizing the embodied and embedded, relational and affective positions of humans is a form of situated knowledge that enhances the singular and collective capacity for both ethical accountability and alternative ways of producing knowledge" (2019, 12).

The act of making legible and learning to read our socio-technical states of unhomeliness leverages our capacity for creativity and human connection, and these processes precisely counteract datafication's domination of the knowledge sphere and control over the knowledge commons. As Braidotti argues, we can "take a distance" from different forms of affiliation, and through this create new affirmative and generative values that form a basis for alternative forms of subjective expression, both as individuals and as groups. This is another way of saying that under conditions of immanence, we need to re-ground the forces of connection and sense-making in local conditions, needs, and ethical orientations. In this way we make datafication legible as a social process of co-production, and thus gain a form of literacy that is productive of human relations.

COMMUNING LITERACY

To commune is to enact. It is to connect, to communicate, to make legible, and to mediate. We produce our world through our ways of communing with one another. Communing data literacy – enacting our ability to mediate – is a means to counteract datafication's symbolic and material power and its ability to assert ontologizing forces. Communing does this by manifesting the epistemic and mediating power of dialogue and collective meaning-making.

Comunalidad is living and being in ways that produce a "we." As Martínez Luna explains, it is an "experiential concept that enables integral, total, natural and common comprehension of making life," (2015, 100) or as Guerrero Osorio explains, "Communality is the verbal predicate of the We. It names its action and not its ontology" (2019, 130). Communality draws our attention to the materialization of perceiving and knowing expressions in and through people's intentional interactions and defends the value of these creative and caring energies in the face of forces that seek to remediate or appropriate them. *Sentipensar*, then, captures the rejection of ontologizing forces that seek to define the channels of appropriation by establishing an "aesthetics of effort" (Cepeda 2017, 19), which is carried out through "the non-pretentious facticity of the encounter with others, listening next to the fire, learning with humility" (17). Thus, *comunalidad digital* is a "counterproposal to capitalist forms of territorial development" (Sursiendo 2022, 26), which demands that we inhabit the complex interlacing of digital and analogue systems in ways that are "conscious of the sociotechnical and environmental impacts they entail" (50). What that looks like in practice will reflect, well, practice, and that means it will be historically, ecologically, and culturally situated, and also dependent on human processes.

Communality emphasizes the caring and creative processes that generate human ties to each other and to the natural world, in sharp contrast to unthinking forms of networking that can be facilitated by datafied processes. To the extent that human connection performs meaning, culture, and knowledge into being, then *comunalidad* and *sentipensar* lay claim to epistemic generative practice. They offer a contrast to the discursive acts of ontological capture performed by corporate actors through the formation of databases and indexation of terrains of being. Communality and feel-think help us differentiate between sustained and substantive efforts to form meaningful human connection, and the shallow affiliations established through datafied processes.

Comunalidad and *sentipensar* are specific concepts created by community members and leaders in contexts of cultural and political struggle. For the Mixe and Zapotec peoples of Oaxaca, *comunalidad* is an expression of resistance to colonial rule. When Martínez Luna says that *comunalidad* prioritizes living and being in ways that produce a "we," he is manifesting a connection between people, land, and history that pushes back on European colonization. He is saying that this connection to the land and the people is enacted through ongoing processes of recognition, sharing, mutuality, orality, remembrance, and hope (Martínez Luna 2015; Guerrero Osorio 2019). This is very much a project of enacting a community and expressing radical communal values, a truth which was asserted most recently at the EZLN's thirtieth anniversary celebrations (Oropeza Alvarez 2024).

Sentipensar, meanwhile, emerges from the "amphibious culture" (meaning land and water culture) of the river-dwellers of the Magdalena River of Colombia. The concept was first documented by Fals Borda (1998) in his ethnographic and participatory action work with Indigenous and campesino people from these regions. The hunters and fishermen he worked with explained that they act both with their hearts and with their minds and that this made them *sentipensantes* (feel-thinkers). They further explained that they remained connected to the culture of living on both the land and the water through feeling and being conscious of their relationship to other people, the land, and nature. Thus Cepeda

(2017) explains *sentipensar* is about making sense *with* rather than making sense *about* – making sense with each other and also with other beings within a context. As in the previous case, *sentipensar* originally served to manifest relations in context, and thus can be mobilized as a powerful expression of community and resistance. Thus, for example, Escobar (2014) uses *sentipensar* to ground his exploration of the eco-communal resistance struggles of Afro-Colombian communities.

We are drawn to these ideas as inspiration for counteracting the atomizing and ontologizing effects of datafication. Communities of convenience organize us in ways that create barriers to forming groups, collectivizing, or expressing our desires in local historical relief. They also organize us in ways that set us adrift, creating the sense that we are unhomed subjects. The symbolic and material power of communities of convenience generate goods, services, and narratives that make it difficult to think outside of these alienating and divisive forces. They make it difficult to understand and overcome the novel subjectivizing forces of the new digital economy and their tendency to engender disconnection, divisiveness, isolation, or apathy. *Comunalidad* and *sentipensar* provide inspiration for making sense of the world through grounded human connection, dialogue, and reflection, and in doing so, they counter the symbolic and material forces of data colonialism and technological alienation.

By communing data literacy, we emphasize processes of encounter, discovery, and meaning-making. Prioritizing living and being in ways that produce connection, empathy, and understanding is an attitude and a form of potential. It may result in the manifestation of projects, collectives, communities, or even resistance movements, and if so, that's wonderful. But if it does not, sitting down to share our observations with others will still make ontologizing forces more legible, and people will still be engaged in practices that counter the isolating logic of communities of convenience.

Doing this work implies exploring the consequence and responsibility of becoming *with* the other, encountering experience and emotion as a means to activate care and relationality in communities of *experience*. In this way, people simultaneously make material their understanding of the experience of living in a datafied society and also actualize the potential for new ways of being, new social connections, and possibly even new actions that confront the displacement of flexibilized and destabilizing systems.

We further argue that to develop literacy among unhomed data subjects, what is needed is to focus on our emotions, narrations, and experiences, and not on what is already abstracted. We focus on getting people to think through mismatches between their needs and desires and the logics of data and information systems. Drawing on Braidotti, we find a way forward in her assertion that "[a] people is always missing and virtual, in that it needs to be actualized and assembled. It is the result of a praxis, a collective engagement to produce different assemblages" (2019, 52). Accordingly, through encounters with the other – and their descriptions of the cracks, nuances, and impacts of living with data – we build a collaborative map of the multiple terrains we inhabit as unhomed data subjects. Through these processes, data subjects can appropriate the epistemic heart of knowledge production and manifest generative spaces of connection.

We also leave open the possibility that participants in data literacy forums will express difference and dissent. We feel that differences of opinion are necessary so long as they remain respectful, open, and engaged. If participants accept how a community of convenience uses their data and organizes social processes, or if only some of them feel motivated to pursue advocacy or social change initiatives, that's also fine. We are not suggesting that people need to choose between community connections and digital services; rather, we are offering strategies to make datafication's mediation of their daily life more legible so that they are better able to navigate its opportunities, demands, and impositions.

Ultimately, we come back to the idea that datafication has enabled new forms of social organization – what we've called communities of convenience – that leave people epistemologically (Fricker 2007) or hermeneutically (Couldry 2023) unhomed. The goal of data literacy in this case must be to give culturally and technologically unsettled subjects flashlights and invite them to shine beams onto each other's individual experiences of digital assembly. By illuminating their social, cultural, and economic reality, they can begin to make sense of how their data is used to organize social processes, produce streams of value, and assert ways of knowing that radically change their human, relational, and communal experience. Once they make these experiences legible, the members of communities of convenience will be better able to support the systems that work for them, and demand changes to the ones that don't. In this way, they are enacting alternative knowledge bases and communing data literacy.

Section Three

—

Techniques to Commune Data Literacy

HOW TO USE THE TECHNIQUES

The techniques presented in this section, which are summarized in table 1, build on the concepts we've developed through real world experiences with data literacy interventions. This includes the idea that we live in digitally mediated spaces that shape how we perceive and know the world we inhabit; that what is real to us is what we experience in these spaces: how they make us feel, and whether they result in well-being and happiness; and that legibility and consciousness come from connecting with others to share our experiences of connecting to these spaces. We connect with each other in order to:

1. Make sense, each from our own situated point of view, of the digital forces that assemble us into platforms, databases, or programs.
2. Become aware of other people's experiences of these spaces and how they are similar to or different from our own experiences.
3. Form empathy for the people we are interacting with, and through this engage in creative processes of becoming different from how we are.

We believe that this work requires spaces to share experiences and engage in dialogue. Without a process of sharing, awareness-building, and empathy, we have little possibility to form points of reference, touchstones, baselines, or benchmarks for assessing our various experiences of being assembled, deciding what is advantageous or problematic about datafication, and deciding how to respond, either individually or as a group. To examine our individual encounters with datafication, we need access to communal perspectives. They will help us develop consciousness of what it means to be entangled with datafied systems and enable us to mediate our experiences otherwise.

Using our jargon, we would say that communities of convenience can't be legible to unhomed data subjects unless and until they begin to share their experiences of these systems with each other. And we can't form literate readings of these systems until they are legible. We would also say that becoming conscious of our experience through sharing and engagement is already an act of counter-power because the logics of communing with each other counteract the atomizing logics of data-mediated affiliation.

The techniques are designed to foreground a contextualized and participatory approach based on the values, needs, and lived experiences of participants. They start by asking participants to share their everyday experiences and then offer ways to make sense of them by mapping imagined positions, discussing values, and exploring feelings. This process allows us to reorient the starting point of literacy from deficit to lived experience.

Each technique offers a theoretical and conceptual grounding followed by step-by-step instructions for workshop activities, and a closing invitation for collective reflection. We've ordered the techniques assuming that a workshop has convened members of a particular community of convenience, but that they are strangers to each other. The goal of the workshop is for participants to share their experiences of that system so that together they can make its effects more legible. And we assume that in the process of generating this multi-angled legibility, they will explore their connections with each other. The first technique, "What's in a Name," doubles as an ice-breaker to help participants get to know each other. The last few techniques provoke reflection about possibilities for change and opportunities for strategic intervention.

Having said this, in practice we fully expect that these techniques will be selected pragmatically and adapted to the needs and circumstances at hand. They should be a starting point for creative engagement, rather than a rule book about how to connect with people. With this in mind, it's helpful to know that each technique starts with a brief explanation of its logic, followed by step-by-step instructions for workshop facilitation, plus an example of how it might be applied. Some people may find it helpful to read things the other way around, starting with the examples followed by the instructions and concluding with the logic of the activity.

Table 1: Overview of data literacy techniques

Activity	Starting Point	Big Questions	Objectives	Type of Activity
1 What's in a name	We have been convened by an infrastructure which positions us as an anonymous and fractured user base.	Who are we as a group of humans? How do we come to know each other as people?	Create potential for the group to engage with alternative logics of community. Shift from being strangers or data points to people with shared experiences.	Sharing names.
2 What data doesn't know	Communities of Convenience are formed on the basis of residuals.	How do residual categories limit us? How do we exceed the categories used by the system?	Recognize our differences and also what unites the group.	Critical reflection about data and information systems.
3 Rights and responsibilities	Communities of Convenience position the technology owner as the steward of the data and centralize control over collective benefits.	Who holds the resources? Who gets to decide? Who benefits? What are my rights?	Identify opportunities to demand certain rights individually or as a group.	Critical analysis of legal and policy frameworks.
4 The value of sharing	Participation in a Community of Convenience is a value proposition, and that proposition may be inequitable.	What is the value proposition at play? What are its advantages and disadvantages?	Identify the value proposition at play and also consider strategies to challenge it.	Mapping exercise.
5 Imagining Data journeys	Communities of Convenience make us part of a socio-material knowledge infrastructure.	How do we interact with / contribute to / make use of this socio-technical knowledge apparatus?	Imagine alternative socio-material knowledge practices that leverage data differently.	Storytelling.
6 Epistemic and hermeneutic justice	We are part of a knowledge project, but face barriers to understanding this project.	What can we know, and what can't we know? What alternative knowledge do we hold?	Identify means to challenge cultural, material, and political expressions of epistemic injustice. Leverage alternative forms of knowledge.	Critical reflection about what we can or cannot know.
7 Citizen celebrities	Communities of Convenience define the parameters of our agency.	What can we negotiate? Why or why not?	Identify individual or collective strategies to build a sense of agency or shift the terms of the negotiation.	Role-playing.
8 Value versus values	Communities of Convenience reflect a set of values.	Does the Community of Affiliation reflect the values of its members? How would it differ if it did?	Develop alternative, grounded values for the group.	Values-based audit.
9 Overflow	Communities of Convenience create social expectations that reify our fractured identity.	How do social expectations (mis)align with personal needs or desires? How does this shape our internal dialogue and external actions?	Identify individual or collective strategies to challenge expectations and shift socio-technical systems towards fulfilling our desires or needs.	Personal reflection.

1
What's in a Name?

It is difficult to imagine the multiple and overlapping communities of convenience we belong to as a result of sharing our personal data with information systems. If we were to convene a gathering of the members of just one of those communities of convenience, it is difficult to imagine who would show up. Once convened, the group's temptation might be to understand our communities of convenience in terms of the socio-technical systems that brought us together – as a community of X-brand loyalty card users or followers of the Y-brand social media feed. However, this is problematic because in doing so we reify technology as our starting point and accept its organizational premise as ground zero for our relationships. We would be "seeing like an infrastructure" (Singh and Jackson 2021) rather than seeing like a community. And we would not be forming a consciousness of who we are as a group of humans – rather, we would be becoming functionally data literate, which is to say better adapted to the socio-technical systems that we have been inserted into.

To materialize knowledge and perceiving communities, we need to come to know each other – according to our particular ways of seeing – both as humans who are inserted into socio-technical processes and as the complex, creative, and social beings who navigate these insertions and form relationships with each other. This technique is designed to help people see that they have been convened by an infrastructure that positions them as a fractured user base; it also aims to help them see each other as a group of humans, and to establish alternative logics as a basis for coming together.

The work of Arturo Escobar offers two useful starting points that unhomed data subjects can use to begin to form community consciousness. The first concept is lexicality. Escobar argues that globalization and modernity impose a "lexicon" of meanings that prioritize "the individual, rationality, efficiency, private property and, of course, the market" (2014, 58). As a result, "it is increasingly difficult to see our connections to the world and to live them as real" (58). We can use this concept to "conjure up the ghost of modernity" (61) in a datafied computer network, as Escobar puts it. That is to say, we can reveal how data and information systems produce patterns of control through the meaning they prioritize, and contrast this with the vocabulary of relationality, communalism, and connection.

This brings us to a second useful concept: *sentipensar*, which Escobar explores in his book *Sentipensar con la Tierra* (2014) based on the work of Fals Borda (1998), who encountered the term in the fishing communities he worked with in rural Colombia. *Sentipensar* is a complicated word to translate, but is often understood to be a combination of thinking and feeling. Cepeda (2017) defines *sentipensar* as "the non-pretentious facticity of the encounter with others, listening, next to the fire, learning with humility … paying attention respectfully to their saying just as respectfully as we read a scientific paper" (17). He goes on to depict *sentipensar* as reasoning-with in contrast to reasoning-about, saying: "the nature of sentipensar requires a connection from the heart" that allows us to "co-reason" or *corazonar. Corazonar* is another neologism, this time linking the word "heart" (*corazon*) to the word "reason" (*razón*), and also the word "hunch" (*corazonada*). In the context of becoming conscious of our communities of convenience, *sentipensar* offers an approach to developing alternative ways of knowing and being that are based in the logics of community rather than the logics of modernity or control. In listening openly and empathetically to each other, we discover who we are as people, and have a better grounding for examining how information systems shape our collective experience of personhood. These practices offer a basis to overcome the fragmentation produced by the infrastructures that organize us into communities of convenience.

THE TECHNIQUE: WHAT'S IN A NAME?

Names, which are often the first thing collected by databases, offer a powerful means to mark the distinction between identification – how we are indexed and organized by a socio-technical system – and identity – who we are as people. For example, for Indigenous communities in the Pacific Northwest, sharing data holds particular cultural significance (Animikii 2019). The act of sharing one's name carries the obligation of interpersonal and communal responsibilities that are part of a specific way of knowing. In short, by knowing who I am, you have taken on a responsibility to care. When information systems reduce names to transactional cues that structure access to goods and services, people who associate naming with being may experience loss, relive the history of colonialism, feel exploited, and sense a lack of control over their own cultural identity and material well-being.

By exploring the contrast between names as identifiers and names as identity, we develop consciousness of how the lexicality of modernity and control manifests in information systems, and we establish the foundations for alternative ways of knowing and being. When we do this work using an attitude of *sentipensar*, then we are already activating these alternative ways of knowing each other as we move towards forming communality. When we talk about what our names mean to us, we are already forming the basis for empathy and community, and we are already engaging in alternative ways of knowing and being.

PROCESS

In this technique, participants are invited to critically reflect on the role of naming in the process of relating to one another and to socio-technical systems. Follow these steps to conduct the activity, and use exercise 1 to guide the process:

- Take a moment to list some of your different names, such as the names you use with friends and family, your professional names, nicknames, or names used in different cultural settings. Be sure to include the name you use when you register for things like loyalty programs, databases, or information systems. Write your list of names in the first column of the worksheet.

- Pick one of your names, and tell the story of that name to someone sitting near you. For example, discuss who gave you this name, its significance, or where it gets used.
- Now fill in columns 2 and 3, writing down who or where each name gets used, and how you feel about this use. For example, if someone close to you calls you by your nickname at a family function, you might feel affection and belonging. Maybe you feel nothing at all, and that's okay. Take a note of this as well.
- Finally, in column 4, consider how each of these names and name-uses orders your relationship to other people (if at all). For example, if you use your legal name for loyalty programs, what does this tell you about the nature of your relationship to these systems and to the other people within them?

Exercise 1 Worksheet for "what's in a name?"

Names	Where do I use this name? Who calls me this?	How does the use of this name make me feel?	How does this name order my relationships to other people?

Example 1 What's in Katherine's name?

Names	Where do I use this name? Who calls me this?	How does the use of this name make me feel?	How does this name order my relationships to other people?
Katherine Reilly	It's my "everyday" name.	This name acts like a kind of shield. It's my "game face."	This name puts me in a transactional or business relationship with people.
Dr Katherine M.A. Reilly	Colleagues at the university, but only for formal occasions or formal correspondence.	Like I'm making believe or pretending; like I'm playing dress up.	This name puts me in a position of authority. I assume a role and a set of responsibilities.
Kath	This is a more familiar name. I might use it to sign off on friendly emails. My husband calls me this.	This makes me feel casually familiar with people – at ease.	This name tells me I am in community, among friends. I feel a sense of care and understanding.
Tiki	My immediate family, especially my two older bothers … oops, I mean brothers.	This name conjures up all sorts of family connections – all of the baggage and all of the bliss.	It's complicated!

COLLECTIVE DISCUSSION

Based on this exercise, discuss the following questions as a group:

- What name do you typically use when you interact with the information system (loyalty program, database, worker platform, etc.) that brought everyone together today? How does the use of this name order your relationships to other people, and how does that make you feel?
- What does this exercise suggest about the effects of information systems and their use of personal data on our connections to the people around us?
- How do you actually want to be known by the people in this group?
- How would you prefer to be connected to the people in this group?

2

WHAT DATA DOESN'T KNOW ABOUT US

Datafied systems attempt to capture what we are (e.g., our age, our gender) and what we feel (e.g., what we like, what makes us angry) and they use this information to organize us in ways that suit the priorities and objectives of systems' designers. Nevertheless, it is clear that our being is never fully captured by these systems – that is, there is always a gap between what our data doubles represent and who we actually are (Lupton 2021). Indeed, scholars such as Star and Bowker (2007) have noted that while socio-technical systems are increasingly mobilized to categorize humanity, it is impossible for all citizens to fully fit those categories. Our identities and experiences often inhabit residual categories that fall outside of formal systems of classification. What is more, databases are often incomplete, reflecting the information that emerges from people's uneven technology uses or behaviours. Finally, it's important to mark a difference between knowing something about someone (their age, for example) and using models to predict (with a greater or lesser degree of reliability) something about someone, such as their preferences or emotions.

Critical data studies scholars note that big data uses categorization in ways that prioritize certain ways of knowing and downplay all other epistemologies, for example, Indigenous ways of knowing or feminist practices of care. These dominant categories are part of the systems of control that order us into communities of convenience, and that undermine our efforts to form alternative community practices. In response, Paola Ricaurte argues that "as data becomes the preferred way of representing knowledge in our time, we must pay attention to the diverse worldviews that come into tension in an unequal world" (2019, 361). Residuality – what does not fit neatly into categories or what is not collected or analyzed – can be read as both a form of violence and a source of resistance to these ordering processes.

Having said this, unhomed data subjects inhabit multiple information systems at the same time, where each of them is emphasizing certain categories of existence, which, in turn, create multiple residual categories. Adapting the words of John Law (2004), as the unhomed data subject engages with multiple socio-technical systems simultaneously, they turn many realities into existence. In this context, a critical examination of what information systems know – and, more importantly, what they don't know – is a useful way to understand the forms of violence and resistance experienced by unhomed data subjects and made possible by communities of convenience. By doing this work, the complex datafied landscape becomes more legible, and datafied subjects gain insights that help them organize their experience. By exploring our different experiences of fitting into or exceeding the system, we can begin to identify potential spaces of overlap and coming together.

THE TECHNIQUE: WHAT DATA DOESN'T KNOW ABOUT US

In this technique, participants are invited to critically examine the types of data that can or cannot be captured about themselves. The activity should not be used to amass personal data; we don't want to replicate datafication, but rather explore how datafication does or does not order our relationships. Participants should be encouraged to think in terms of categories such as "name" or "age" and not specifics like "Mary, aged 38." This activity has three steps:

- Consider an information system that the members of the group belong to. Spend some time thinking about how this system works: what is this system designed to do? What types of data does this information system collect? Why?
- Using exercise 2, write down aspects of yourself that you feel are being collected by this system. Sort these categories into information about who you are, what you do, and what you feel.

- Do the same thing again, using the worksheet to write down the aspects of yourself that cannot be collected as data points, or that would be poorly reflected in data points.
- Finally, consider what this information system might get wrong, or how it might fail, based on what it can or can't know about you.

Exercise 2 Worksheet for "what data doesn't know about us"

	Who I am	What I do	What I feel
Can be / is collected			
Can't be collected / Difficult to collect			
What they could get wrong			

THE EXAMPLE OF CREDIT RISK MODELS AND SUPERAPPS

SuperApps are digital applications that bring together several digital services in one place. A Latin American example is Rappi, which combines delivery, e-commerce, digital wallets (e-banking), and gaming. In a 2021 study, Roa et al. (2021) used Rappi data to explore new ways of assessing credit scores. They concluded that a combination of financial data and platform data improves a lender's ability to predict if someone will default on a loan. "Payment errors" that happen when someone is ordering delivery through Rappi (such as declined credit cards) were found to be the best indicator of the probability that someone would default on a loan. Roa et al. concluded that Rappi could use this information to determine if someone was a poor candidate for a loan. SuperApps such as Rappi offer an interesting opportunity to think about what an information system can know about us.

Example 2 Analysis of the credit risk model used by Rappi

	Who I am	**What I do**	**What I feel**
Can be / Is collected	Rappi knows its users' gender; age; tenure on the app; country and city of residence; most used address; number of different addresses; number of registered credit cards and their brands; phone brand and operating system; number of different phones used.	Rappi knows its users' preferred payment methods; orders; order value; cancelled orders; refunds; payment methods; shopping patterns; tipping patterns; transportation patterns; number of transactions and amount spent via debit or credit; money transfers; cash withdrawals.	Rappi tracks preferences, such as preferred vehicle for transportation, and feedback on services offered, for example through ratings, tipping behaviour, or repeat service.
Can't be collected / Difficult to collect	They don't know my beliefs, motivations, and values. They can *predict* these based on models and analysis, but they don't know for sure. They also don't know me as a member of a family, or as someone with commitments to my community.	They can only know what I transact through their system. They may be able to *guess* other things I do based on the information they gather.	They can know my ratings, but they can only *predict* my feelings based on models that will always contain biases. They probably only want to know the feelings that are relevant to their needs – not all of my feelings. And they don't look at my feelings in relation to others.
What they could get wrong	If I have multiple accounts, if I share an account with friends or family, if I'm an infrequent user, or if I'm a member of a marginalized or minority group, then their picture of who I am is more likely to be incomplete or incorrect.	They may misinterpret my activities if I do something unexpected: if I'm in a moment of transition (for example between jobs), make one-off purchases (such as a gift), or change my spending habits (for example to save up for something).	The more varied the pool of people, situations, and feelings the data is working with, the less accurate the predictions become. Prediction requires patterns. People's feelings, emotions, or behaviours are not always easy to predict.

COLLECTIVE REFLECTION

This technique reveals the potential and limitations of datafication as a system of organization and control. It invites us to identify the forms of harm that are enacted when we are categorized by or excluded from datafication processes, and it also encourages us to reflect on how we exceed categorizations and how this can create spaces to resist, engage in alternative practices, or develop alternative systems. With this in mind, reflect with your group on the following questions:

- What can be known about users through datafication, and what can't be known?
- What data do you wish were collected that isn't? What data is collected, but you wish it weren't?
- How might you be harmed by the data collected? How could you be harmed by the data that was not collected?
- How does datafication's attempt to "know us" shape our individual and collective experience?
- How do our individual experiences differ and how are they similar? Do our common experiences offer a basis for connection, relationality, and care? How so, or why not?

3

RIGHTS AND RESPONSIBILITIES

Communities of convenience often position the owners of technology as the stewards of our personal data, and systems of rights and responsibility are frequently designed to uphold this arrangement. The legal frameworks designed to protect data subjects are also designed to respond to their concerns as individuals. For example, personal data protection laws offer a legal basis for citizens to know, adjust, or challenge how their data is being used by companies and the state (Ausloos and Dewitte 2018). This legal protection is designed to enable data collection by technology owners, but leaves individuals responsible for managing their many different experiences of datafication. Unfortunately, this means that data protections may contribute to our experience of being unhomed data subjects, and this can make us feel powerless to challenge the experiences of community that result from datafication.

This is a challenging reality of the data landscape; however, the process of thinking about and identifying data protection tools can still offer a valuable opportunity to develop consciousness about the context for datafication. Sara Ahmed argues that becoming aware of, or using, whatever tools are made available by the powers that be allows people to name, explain, and respond to power imbalances. Thus, even if efforts to make datafication legible do not achieve the ultimate goal of changing data collection protocols, "it would still have effects" (Ahmed 2021, 59) by providing people with a means to "read" and challenge the socio-technical regime we inhabit. Certainly, this is a necessary first step toward using collective power to demand changes in patterns of data use, collection, and circulation and their regulation. By exploring the tools available to citizens – whether in the form of laws or technological solutions – unhomed data subjects are better able to understand and address power imbalances between ourselves and the systems that collect and analyze our data.

THE TECHNIQUE: RIGHTS AND RESPONSIBILITIES

Drawing from the history of social audits (de Lima Pinel, Cosenza, and Macarulla 2015) and our work on citizen data audits (Reilly 2020; Reilly and E. Morales 2023), this technique seeks to provide individuals and communities with opportunities to identify, become familiar with, and apply data protection laws available to them in their local context. It may be helpful to locate relevant documents ahead of time to facilitate this process, and/or to invite local digital rights experts to participate in these activities. In our experience, people have many technical questions about their legal rights and official processes!

It is important to note that while data protection mechanisms are more or less taken for granted in countries of the Global North, such as Canada (e.g., Hilts and Parsons 2014), they are not always available to citizens in Latin America. For instance, while Peru has a mechanism to demand companies' transparency and accountability around their use of personal data (Reilly 2022), as of the publication of this volume Paraguay has yet to establish a personal data protection law (TEDIC 2023). If there are no data protection laws in your context, you might find it helpful to consider one of the other techniques presented in this chapter. For instance, Citizen Celebrities asks participants to think of themselves as the superstar producers of valuable cultural products, and has them imagine the contract they would negotiate with a socio-technical system on the basis of their value and importance.

In what we describe below, participants are simply asked to become familiar with legal frameworks. We have also gotten community members to apply legal provisions, which is a more in-depth way to become familiar with their workings. The latter is outlined in *Son Mis Datos: Guía Metodológico* (Reilly et al. 2020) and discussed in Reilly (2022). Much can be gained from having data subjects submit personal access to information requests through personal data protection laws;

however, based on our experience, this tends to position people as part of the state apparatus and overall system of data appropriation. So in this activity we have focused instead on exploring how legal provisions set up the roles and relations between different actors within the data assembly, and then invite participants to reflect on what this means for their experience of being assembled or unhomed. The goal is to begin to identify how data assemblages create the conditions for communities of convenience, and possibly also opportunities to come together around collective responses.

PROCESS

To conduct this technique, gather in small groups and choose a particular interaction with a socio-technical system (such as shopping at a pharmacy or registering for a social media platform).

- Identify relevant data laws or policies that govern that particular service or company. In many cases this will be a personal data protection law; however, there are often specialized laws for certain sectors such as health and education. You can also check the terms of service agreement you signed on to when you became a member of a community of convenience such as a platform, database, loyalty program, or similar.
- Identify the actors involved in these policies. For example, is this law implemented by a Personal Data Protection Agency? Does the company in question have a Personal Data Protection officer? Does the law define data holders in any particular way – as "users," for example?
- Using exercise 3, write down each actor's rights and responsibilities.
- Finally, write down the process through which each actor can exercise their rights.

Exercise 3 Worksheet for "rights and responsibilities"

Actors	Rights	Responsibilities	Process

Example 3 How Canada's *Personal Information Protection and Electronic Documents Act (PIPEDA)* shapes rights and responsibilities

Actors	Rights	Responsibilities	Process
Private-sector actors who collect personally identifying information for commercial transactions.	"An organization may collect, use or disclose personal information only for purposes that a reasonable person would consider are appropriate in the circumstances."	Follow the ten fair information principles covered by PIPEDA. Appoint a privacy officer, and communicate that person's name or title internally and externally.	Private actors must obtain consent to collect, use, or disclose personal information.
Corporate Privacy Officer		Communicate the company's personal data management plan. Facilitate individual access to personal data held by the company. Review and respond to challenges to corporate compliance.	Usually by email or webform. The officer must reply to inquiries within thirty days.
Individuals / People	People have the right to access the personal information held by an organization. They also have the right to challenge its accuracy.	In some countries, citizens are required by law to ensure that the information they share is accurate; however, this is not the case in Canada.	Contact the Corporate Privacy Officer – usually by email. Escalate to the Privacy Commissioner of Canada via the website.
Privacy Commissioner of Canada	Audit companies and review challenges to corporate compliance with the law.	Audit companies and review challenges to corporate compliance with the law.	

COLLECTIVE REFLECTION

Once you've mapped out the different rights and responsibilities made possible by these legal or policy frameworks, discuss the following questions:

- What rights do these legal frameworks or policies give you, given your role? What rights do you wish you had?
- How likely are you to use the tools and processes made available by this system? How could they be made more accessible? What tools or processes do you wish you had access to?
- Who benefits most from these frameworks or policies? Who benefits least?
- What do you notice about how power is organized through these legal frameworks? Who do they empower most? How so?
- What can be done as an individual through these frameworks or policies? What could be done if members of a community of convenience came together as a group to leverage these processes?
- What would need to change in order for people to work collectively to address power imbalances with data holders?

4

THE VALUE OF SHARING

Sharing our data with a community of convenience is a value proposition, and the different members of a digital network will experience that value proposition differently. A value proposition is the individual and collective experience of gain or loss that results from entering into an exchange. Ideally, the parties to an exchange ask themselves whether the benefits gained from a transaction are worth the costs incurred. However, people are often unaware of the value proposition at work where personal data is concerned, which is a significant problem. Also, people typically pursue this negotiation as individuals rather than groups, and this leaves uninformed, vulnerable, or marginalized individuals at risk of experiencing a disproportionate share of value asymmetries.

It is entirely possible for value propositions to benefit both parties. For example, if you were looking to buy a new car, you might be happy to disclose your intentions if it meant receiving useful information. In other cases, there are efforts to compensate members for their data. For example, Optimity (https://www.myoptimity.com) is an app that allows insurance companies, consumer brands, and employers to fulfill their corporate social responsibility agenda by distributing coupons and points to people who achieve fitness goals. Of course, only those who are able to use this system have access to the compensation provided.

Other scenarios are crafted to exploit fundamental value asymmetries (Dodds 2017b). Every time we fill out a reCAPTCHA, we are enabling artificial intelligence; for example, when we identify the crosswalks in photographs, we are teaching driverless cars how to "see." From the user's point of view, this task is the price we must pay to access the service we want to use. But the driverless car technology will eventually be sold to us, even though we helped produce the data set used to train the car. Ekbia and Nardi (2017) call this heteromation: automated socio-technical processes that mobilize individual users to generate economic value for companies. Using an automated checkout is also a form of heteromation, in that we become part of the system that accumulates value for food distributors, but do not receive any compensation for this. Clearly, it is important to become conscious of the value propositions that shape our experience of information systems. Among other things, they structure our social relations and contribute to the forces that determine whether and how we can come together as groups. They also have important implications for how datafication shapes equity and structure inequality.

THE TECHNIQUE: THE VALUE OF SHARING

The goal of this activity is to help the members of communities of convenience become more conscious of the value propositions they entered into through sharing their personal data. This activity may also help members of an information system identify individual or collective projects to reshape the value proposition at work. For example, Mark Kear (2017) explores how the marginalized members of a peer-lending program generated localized data in an effort to influence their credit scores so they'd have better or more favorable access to loans.

PROCESS

This activity builds on a framework for mapping data flows created by Dodds (2017a), but where Dodds is mapping a data ecosystem according to the roles of different data handlers and the flows of value produced by the network, here we centre the experience of personal data holders and prioritize mapping different actors and the value propositions they've entered into. As a result, the focus of this exercise is to understand how value flows structure our social relations.

Start by sharing exercise 4.1, which is a handout called "Mapping Data Value Propositions" (designed by Jihyun Park). As you work through the activity, it might be helpful to move between small group discussion

Exercise 4.1 Mapping Data Value Propositions

Actors and their Roles

Convenor	The owner or provider of the platform, network, or program. They govern the terms of participation.
Complementary Actor	Services or supports that enable the work of the convenor. For example, digital payments services enable the work of platform marketplaces.
Members	Individuals, users, customers or participants who join the platform.
Provider	Actors that provide services, sell goods, or deliver support through the network. These might include mentors, manufacturers, developers, educational programs, or the like.
Intermediaries	Actors that convene communities, working groups, sub-networks, etc., within a larger platform, network, or program.
Advertisers	Entities that leverage the value proposition created by the network to promote services, goods or supports.
Regulators	Governments or policy makers that set the conditions for the tangible or intangible values that can flow through between the actors in the space.

Legend of Mapping Icons

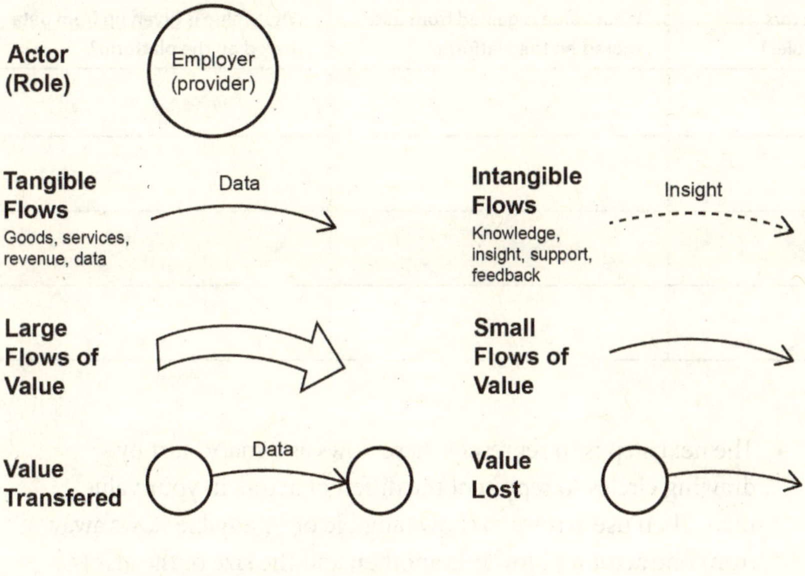

Based on Leigh Dodds. 2017. "Mapping Data Ecosystems."
https://blog.ldodds.com/2017/03/13/some-tips-for-open-data-ecosystem-mapping/

and plenary reflection. For example, after brainstorming in small groups for step one, the full group could gather to discuss the list of actors and types of value flows they will map during step two.

- Using exercise 4.2, identify the main actors in your community of convenience, and add their roles in brackets. Note that we've set up these tables assuming that this analysis will be done by members of these spaces (the individuals, users, customers, or participants who join the platform) but the exercise can also be done from the point of view of other actors such as providers or advertisers. In that case, it's simply a case of being clear about who is asking the questions.
- Consider what you gain from sharing data with each actor, and consider what they gain from gathering your data. Consider both the tangible and intangible value of data flows, as well as the magnitude of the flows. As Dodds (2017a) notes, tangible things are goods, services, revenues, or data itself. Intangible things might be knowledge, feedback, endorsements, or similar.

Exercise 4.2 Worksheet for "the value of sharing"

Actors (Roles)	What value is gained from data shared on the platform?	What value is given up from data shared on the platform?

- The next step is to represent these flows as a map. Start by drawing circles to represent the different actors in your value map. Then use arrows to show tangible or intangible flows away from one actor and towards another, and the size of the arrow (fat or skinny) to represent the magnitude of the value flow. It is entirely possible for systems to lose value, as when resources are poorly invested, lost, or stolen. This can be captured by an arrow that points away from actors. If your list of actors is very long, you might decide to focus on relevant, interesting, or problematic relationships and value propositions.

EXAMPLE: AN EXPLORATION OF THE VALUE PROPOSITION OF DATA FLOWS IN LINKEDIN, FROM THE POINT OF VIEW OF A WORKER

LinkedIn is a social media platform for workers and businesses, and it is also a labour-market ecosystem that enables professional networking and circulates information, services, and training of relevance to workers. LinkedIn also generates data about the actors on its platform through a service called LinkedIn Talent Insights, which, according to the website, provides "[r]eal-time data to hire and manage your workforce" (LinkedIn 2024). Here we offer some examples of how LinkedIn members might visualize value propositions, and from this, make sense of the relationship between data-sharing and social relations.

Example 4.1 Analysis of the value of sharing data on LinkedIn

Actors (Roles)	**What value is gained from data shared on the platform?**	**What value is given up from data shared on the platform?**
Workers (Members)	Workers can search job ads and research companies that are hiring. (Tangible: information about job opportunities) Workers can share an authenticated resume. (Intangible: credibility; tangible: discoverability)	By participating in LinkedIn, workers share data about the labour marketplace. (Intangible: insights)Their participation creates a competitive marketplace for employment, so they lose leverage to negotiate wages and working conditions. (Intangible: negotiating power)
LinkedIn (Convenor)	LinkedIn gains insights about the labour marketplace which they can sell back to employers. (Intangible: insights; tangible: data sales)	None.
Employers (Providers)	Employers gain information and access to services that give them greater control over hiring processes. (Tangible: information to define jobs; wage information)	Employers share insights about their industry with LinkedIn. (Intangible: insights)

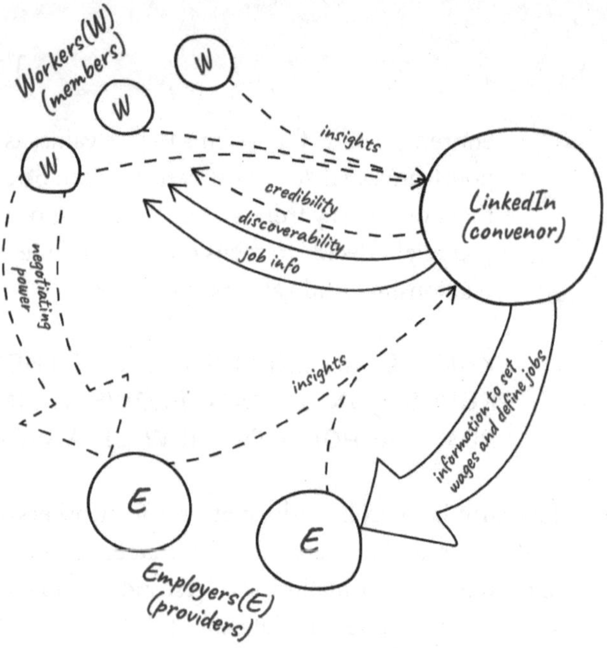

Example 4.2
Map of LinkedIn's value proposition

COLLECTIVE REFLECTION

Based on the previous activity, reflect on the following items:

- Are you able to identify all the actors who influence the data value proposition? Are you able to identify all the value flows resulting from the circulation of data? Reflect on what prevents you from knowing more about the data value proposition at work in your community of convenience.
- Based on the work you did, are the benefits worth the costs? How so or why not?
- How do legal provisions, systems designs, economic incentives, or the like influence your decision to enter into this data value proposition?
- Where does value from data circulation accumulate in this community of convenience? Are you happy with the overall flow of value from this system? Why or why not?
- What can you do as an individual to improve the value proposition at work within a community of convenience? What could you do as a group? Think about strategies that can be employed internally to the socio-technical system you interface with, such as deliberate forms of data-sharing (Kear 2017), and also strategies external to the system, such as forming a user coalition to defend your rights.

5

IMAGINING DATA JOURNEYS

Data goes on journeys (Bates, Lin, and Goodale 2016). It emerges through carefully curated engagements with socio-technical systems (the registration page for a loyalty program; an online transaction with an e-tailer), passes through specific devices (a cell phone; a computer), travels across cables, is stored in data centres (often located in another country), and is analyzed according to particular agendas and goals. Data flows through infrastructures that have been designed to curate and shape meaning in particular ways, and these infrastructures are all around us (Bates, Lin, and Goodale 2016; Akbari 2020). This means that when we become part of a community of convenience, we become part of a social and material knowledge infrastructure that draws on our data-essences to enact culture, organize work flows, make decisions, produce innovations, or generate new insights. Our data and its journeys are epistemic: they produce knowledge, and when we share data we become part of that epistemic infrastructure.

Data's journeys are driven and shaped by the actors who use and manage the data – actors who are embedded in specific networks of power. In this context, it is not surprising that, in general, the actors and places that dominate data journeys are located in the Global North (Ricaurte 2019). In this sense, to think about data's journeys is to reflect on the infrastructures that sustain contemporary relations of power, and how they shape experiences of control, colonization, resistance, and connection (Estrada and Lehuedé 2022). By narrating data's journeys, unhomed data subjects can "read" communities of convenience and data assemblies into legibility. And in doing this, they also reclaim power over the stories that get told with their data, and a measure of control over the knowledge infrastructure.

THE TECHNIQUE: IMAGINING DATA JOURNEYS

This technique mobilizes our imaginaries of data journeys to make legible and re-narrate the knowledge apparatus that is produced by data's trajectories. Often data's real journeys happen in a black box so we can't really know how data flows, where it is stored, or how it is analyzed. But we invite participants to imagine data's flows based on the people, institutions, and discourses that fall outside of the black box. The activity asks participants to think about their community of convenience and the data journeys it organizes as a knowledge apparatus. If your loyalty program were a library, what kinds of knowledge would you find there? What role do we play in producing that knowledge that circulates there? How could our data be organized differently, and what alternative forms of knowledge would result?

PROCESS

Follow these steps to create an imaginary map that represents the route that your data takes once it is collected and systematized.

- Invite participants to reflect on a significant moment of data-sharing. What data were they asked to share, and what was the context for this? Have them share their experience with a partner or in small groups.
- Have them draw or describe an imaginary map of their data's journey after it was shared. Encourage participants to be creative. For example, they could give their data a persona and send it on a trip that passes through several locations. Have them think creatively about locations. These can be geographical places, but also technological destinations such as servers, platforms, or spreadsheets, or even the minds and mouths of people who speak through our data.
- Also, ask participants to think about the protagonists on this data journey: who uses and handles their data, how, and

for what purpose? Have them think about how the data is transformed through these interactions, and also how its value is transformed. Again, encourage participants to be creative. Perhaps the protagonists are robots, algorithms, or AI.

- Finally, have participants reflect on the ultimate destination of these data journeys: What kind of knowledge got produced through these processes? After all that movement and transformation, what could be known, done, created, or produced out of their data?
- Have each participant share their narrative with other members of the group and invite listeners to reflect on actors or routes they hadn't thought of. Also, have them discuss the types of knowledge systems encompassed by their community of convenience: what does our data allow the system to know? And by extension, what does this allow the community to know about itself?

Example 5.1 Corporate Database Management

Some journalists in Peru signed up for an account with a multinational company that operated satellite TV services in their country. When they registered for the service, they accidentally mistyped a digit in their national ID number. All the rest of their data, including their phone number, was correct. After registering, they started getting promotional text messages on their phone that were addressed to someone else. Let's call her Sra. Fernandez. This allowed them to see which companies were buying phone numbers from the multinational for marketing purposes. They also guessed that the multinational was using national ID numbers to index their databases. Because someone else's ID number was being used to index their account, they could also know that person's name, gender, and other information about them. This suggested that the company's information system was not very secure and was exposing Sra. Fernandez to a data breach. Through their experiences, they were able to piece together an image of how data circulated through the systems created by this company, and also work out the effects that this was having in the community. For example, they figured out that the data was being sold to third parties who were using it to circulate advertising via text messages.

Example 5.2 Imagining My Data's Trip to California

Last week I was cleaning for a lady in Vitacura (a rich neighbourhood in Santiago, Chile). The weather was terrible when I finished, so I didn't feel like walking to the subway. And anyway, I needed to cut across town to get to my next job. My niece had installed a taxi app on my phone, so I decided to try it out. I never use taxis because they are so expensive, but I got a discount because it was my first time. After that they started sending me advertisements for food delivery. I figure my data must have gone on a big trip to California, and while it was there it got a makeover and came back to Chile as a wealthy woman who takes taxis and orders meals from restaurants! My data went to California, but I didn't. I would love to go to California one day.

COLLECTIVE REFLECTION

It is often impossible to fully comprehend our data's journeys. Most of the processes and flows that occur within digital systems are hidden and inaccessible to users. Nevertheless, by imagining data flows we learn a lot about how data is understood and valued within our communities, and this creates a basis for epistemic justice (Akbari 2020). By narrating our data's journey as we imagine it to be, we open the possibility of reimagining it differently. With this in mind, reflect on the following questions:

- What kinds of stories did the group produce, and what does this suggest about our collective experience of datafication?
- What type of knowledge project do we contribute to when we share our data in a particular information system or community of convenience?
- What would you change about your data's journey? What type of knowledge would you rather see produced on the basis of the data that you share?

6

EPISTEMIC AND HERMENEUTIC JUSTICE

When we share our personal data in the digital economy we become part of an epistemic project. Consider an author who is writing a work of fiction. In this literary project the author will gather information and generate ideas so that they can construct a story. They will write the story with a particular audience in mind, such as horror enthusiasts or romance lovers. Some of the information they gather will include data about reader preferences, and the author might use this data to orient the narrative of the book. This will help ensure the success of their publication, because it will reflect the preferences of book-buyers. Similarly, the managers of loyalty programs collect data about users so that they can "author" a marketing "story" and consumer experience (i.e., brand) that attracts and maintains a consumer base. We participate in these projects according to the terms of the author's project, but we don't just participate as audiences, we also participate as data providers.

The content of these "stories" has ideological weight – that is, it influences how we understand ourselves and the world around us. Indeed, drawing on the work of Walter Benjamin (1969) and Jesús Martín-Barbero (1993), the data economy constitutes a *sensorium* – a regime of perception or dominant way of knowing and being. Media and technology help to constitute this regime by performing particular types of stories, experiences, and understandings. But as minority partners in communities of convenience, we face many barriers to understanding the part we play in constructing these stories, or how these stories shape our own self-perception, beliefs, and preferences. This results in epistemic and hermeneutic injustice.

Epistemic injustice broadly refers to inequality in what can be known, while hermeneutic injustice refers to barriers that prevent us from understanding or expressing our own experience (Fricker 2007). Because it is difficult to know how our personal data is used, or how information systems construct our consumer experiences, we experience these ideational forms of injustice when we participate in communities of convenience. This leaves us ill-equipped to make decisions about our participation or respond in proactive ways. Without a clear understanding of how our participation shapes our experience of information systems, it is difficult to understand our experience as unhomed data subjects and to move towards alternatives.

THE TECHNIQUE: EPISTEMIC AND HERMENEUTIC JUSTICE

This technique focuses on identifying, naming, and exploring the epistemic injustices we experience through our participation in communities of convenience. It builds on Nancy Fraser's three-part definition of injustice (1995; 1998) which includes: 1) cultural-symbolic injustices that result from not being recognized (for example, as a woman or as Indigenous); 2) socio-economic injustices that result from an unfair redistribution of material resources; and 3) political injustices that result from unequal or inappropriate representation in decision-making. This framework can be applied to anything, but in this case we are using it to think through ideational or epistemic injustices. For example, information systems have been widely shown to reproduce racism in how they recognize different bodies. Melanin-rich skin tones were not originally indexed in facial recognition systems, for example, so the technology could not recognize or do justice to Black people (Najibi 2020). Socio-technical systems also have significant implications for the (re)distribution of material resources, as when people's geolocation is used to determine whether they will get access to a loan (Scassa 2018). Finally, the mere fact that it is difficult to know how our data is used means that we can't adequately represent our interests. By analyzing epistemic justice in terms of recognition, redistribution, and representation, we make the data economy more legible, and also uncover how our data-sharing is used to reproduce this system.

One last note. You don't need to be an expert in computers and technology to participate in these discussions. Indeed, the idea that we need to be experts to participate in the conversation itself reproduces epistemic injustices. Martín-Barbero's work shows us that in any given regime of perception, what can be known far exceeds the story being told by the dominant epistemic project. People are creative in responding to media and information systems, and their ways of thinking often challenge mainstream stories. There is value in nurturing these alternative ways of knowing because they guide us towards alternative regimes of perception and allow for different ways of being in the world.

PROCESS

Individually or in small groups, select an interaction with a particular socio-technical system. For example, you might think about the last time you made an online purchase, or reflect on an activity you engage in as part of the community of convenience that brought you to this discussion.

- Working with your group, complete exercise 6.1, responding to what we can and cannot know about how our data is collected, analyzed, and circulated.
- Now, using exercise 6.2, identify the forms of cultural and symbolic injustice you notice.

Exercise 6.1 Worksheet to identify what can or can't be known about data flows

We can know / It's easy to know	We can't know / It's difficult to know

Exercise 6.2 Worksheet to think about how data flows shape epistemic justice

Cultural/symbolic: Is my identity fully and equitably recognized? How might this affect me?	Socio-economic: How does data use influence the (re)distribution of resources in this system?	Political: Do I have sufficient knowledge about how my data is being used to represent my interests?

Example 6.1 Banking and Authentication Checks

When customers sign up for a bank account, they are often asked to consent to data-sharing that is used to authenticate their identity and assess their eligibility for products or services. The customer will be asked to provide all sorts of information about themselves, from identification, to credit checks, to education or work histories, to details about net worth. When we ask why the bank needs all this data, we might be told that our data is used to ensure that our money is secure, or that it will be used to enhance service provision.

We know what data we are providing to the bank; however, we don't really know how authentication unfolds, how eligibility is assessed, or how any of this information is used as a basis to profile customers for advertising or access to programs, services, or offers. These processes raise concerns about the possibility that gender identity, ethnicity, the neighbourhood we live in, or other markers may be used in ways that create bias. Meanwhile, the way the bank handles our data will have direct implications for our material well-being because it affects how our money is handled, the fees we pay, or the benefits we receive. However, it is very difficult to articulate these concerns or represent our self-interest to the bank, because we don't actually know how our data is used in authentication and assessment processes.

The inequity between what the bank knows and what we know as customers creates the potential for larger socio-economic injustices, such as systemic racism and inequality. At an epistemic and hermeneutic level, the imbalance in what we can know creates room for the banking industry to dominate the story that gets told about personal data.

Example 6.2 Ride-Sharing and Algorithmic Wage Discrimination

In a 2023 paper, Dubal (2023) presented research which demonstrated that digital variable pay practices allow platform companies to engage in algorithmic wage discrimination. Building on this effort, a 2024 investigative report by the media outlet More Perfect Union (2024) demonstrated how this might work in an experiment carried out with Uber and Lyft drivers. They had seven drivers activate screen capture software, open their ride-hailing app, and place their cellphones on a table next to each other. Then they monitored how much each driver was offered for the same fare. As the host Eric Gardner recounts, "We found that Uber offered the same rides 46 times to multiple drivers. Sixty-three percent of the time, one of the drivers was offered a bit less money for the exact same ride. Lyft was even worse ... We saw differences of up to three to four dollars after bonuses." It is unclear how ride-hailing algorithms calculate the cost of a fare, leaving drivers to wonder how their ratings or acceptance/cancellation rates are interpreted by the system. Drivers are repeatedly told that they should think of themselves as independent contractors, and that these ride-sharing jobs will give them greater flexibility and control, but the lived experience of wage discrimination calls this into question. The resulting unequal treatment of workers has implications for earnings, but without knowing exactly how the pricing algorithm works, it is very difficult for them to represent their interests in the ride-hailing marketplace.

COLLECTIVE REFLECTION

We do not need to become data scientists to understand the epistemic injustice we sense when interacting with information systems. Instead, we can focus on what we can or can't know, and how that enables the types of stories we get told. Our experiences have epistemic weight, and they can be used to counter the dominance of particular ways of knowing or being. With this in mind, conclude this activity by having the group discuss the following questions.

- What can we know and what can't we know, given how this community of convenience is organized?

- What kinds of injustices result from these patterns of knowing or not knowing?
- What types of stories are made possible by this pattern of knowing?
- Why is the regime of perception organized in this way? Who does it benefit?
- How could we get better knowledge about how our data is being used? How would the stories that are told about personal data use differ in this case?
- How can we challenge these injustices, either individually or in groups?

7

CITIZEN CELEBRITIES

Critical media scholars note that many people feel powerless to shape the future of technological development and find participation in datafied spaces unavoidable (Markham 2021b). For unhomed data subjects, it is difficult to (re)imagine a future where we can productively inhabit and appropriate the processes of cultural transformation, particularly in the face of powerful algorithms and deterministic discourses (Siles, Gómez-Cruz, and Ricaurte 2023). We struggle to express agency in how socio-technical systems are designed, implemented, and mobilized.

In this context, we need ways to imagine ourselves beyond the established course of technological development. We need spaces that allow us to experience and express our agency and that help us see how our experiences could be different. Considering the need to recentre agency, this technique helps participants imagine new social contracts with socio-technical systems.

THE TECHNIQUE: CITIZEN CELEBRITIES

This technique is inspired by Colombian communications scholar Ómar Rincón. Rincón's work continues the trajectory of Jesús Martín-Barbero, who famously theorized how people engage with, make sense of, and appropriate media (Martín-Barbero 1993). But while Martín-Barbero celebrated people's interpretive agency in the age of broadcast media, Rincón worries that social media has produced a zombie society: a society in which people are disconnected from life and from each other

(Rincón 2018). He offers the concept of citizen celebrity as a way to bring the zombie back to life. As he says, "We all have the right to be on screen, but according to our own aesthetics, expression, affective force and narrative, and I call that celebrity citizenship" (76).

The citizen celebrity is a person who is conscious of how their relationship with the world is mediated by datafication. This technique asks an unhomed data subject to recentre themselves as the subjects of their own narrative, to re-negotiate their terms of engagement with information systems, and to make explicit what they like (or do not like) about the datafication processes they engage with in their everyday lives.

PROCESS

In this activity, participants are invited to think of themselves as a celebrity who has legal representation and an agent. They are asked to mobilize their position and resources to negotiate the terms of their engagement with an information system. The activity takes place in three moments.

First moment: Ask the participants to consider specific moments of data-sharing and discuss them with a neighbour. For example, they might share the process of registering for a social media platform, or the experience of passing through security at an airport.

- Under what conditions can the company collect your data?
- In what situations is the company not allowed to collect your data?
- Are there any specific pieces of data that you would or would not allow the company to collect or analyze?

Second moment: Ask participants to imagine that they are meeting with their lawyer and agent to think through their data demands. The cues below can help them think about the terms they wish to negotiate.

- What are acceptable or unacceptable uses of your data?
- What do you want the experience of datafication to be like? What kind of aesthetics do you prefer, how do you want to be able to express yourself, how should you feel, and what kind of narrative do you want to belong to?

Third moment: Have the participants role-play a meeting with the manager of the company or organization who wants access to their data. Have them state their value to the company and express their demands.

Example 7 Negotiating with the *New York Times* Games Application

Written by Dr Rachel Horst

Me: I'm thinking of subscribing to your service, *The New York Times*, so that I can use your app on my iPhone. I would like to read the news, of course, but I also want to use your recipe app for a dinner party I have coming up. And the games! Spelling Bee is my favourite! Now, I know you collect abundant forms of direct and indirect data on your users. Before I agree to sign on, I wish to negotiate the terms of your collection of my data. I have my lawyer here to help negotiate those terms.

NYTimes: Of course, we really want your patronage and are willing to negotiate the terms to the best of our ability.

Lawyer: First of all … the Spelling Bee game … Unless my client achieves a genius score, her game data is not to be collected.

NYTimes: But if we only collected top scores, then eventually we would need to make the game more difficult. No one will want to play if *everyone* can win!

Lawyer: This one is non-negotiable. My client has a reputation to maintain.

NYTimes: I suppose we can make an exception here ...

Lawyer: Good, now about the recipes app. How is this data used?

NYTimes: Well, her searches are stored. And we use this data to tailor her experience, along with the kinds of marketing and advertising she will be shown.

Lawyer: My client is a vegetarian.

NYTimes: Not according to her data … She was just looking at a recipe for brisket the other night.

Lawyer: It's for the dinner party. She will prepare meat for her family and friends, but won't eat it herself. It's complicated. Her family is very resistant to eating vegetarian. But my client does not wish to support the industrial meat complex any more than she absolutely has to. Furthermore, she wishes not to have meat products or meat-based recipes promoted to her. So if we can clarify in her data that she is a vegetarian and only reluctantly cooks meat, that would be ideal. Her data needs to convey these nuances.

NYTimes: We will note that in the database.

Lawyer: Now, in terms of her news reading habits. First, my client is okay with you occasionally suggesting stories to her on topics that interest her. She finds this feature enjoyable and timesaving. However, my client also wishes to diversify her reading and not have a siloed news experience, so if you could remind her of areas in the world she is neglecting, or trends she should be more knowledgeable about? She wishes to have a well-rounded and diversified news experience.

NYTimes: Okay, this may require us to create a unique algorithm, just for your client.

Lawyer: This brings us to another point.

NYTimes: Yes?

Lawyer: My client reserves the right to change her preferences at any moment.

NYTimes: Of course, with the understanding that once her data has been shared, it cannot be unshared.

Lawyer: Hence our desire for extreme caution and clarity.

COLLECTIVE REFLECTION

Unhomed data subjects typically have little power to negotiate the terms of data collection and analysis, and as a result, companies and governments can impose constraints on our aesthetic, expressive, affective, and narrative experience (Rincón 2018). By centring agency, this activity produces a critical encounter between unhomed data subjects and the communities of convenience that organize them. To wrap up the activity, consider the following questions:

- What prevents us from negotiating the conditions under which companies can collect and analyze our data?
- If you could have a negotiation with companies, what conditions would you prioritize regarding data collection and analysis?
- How likely is it that you could negotiate these terms on your own? If you worked with a group of people, how would your experience of negotiation change?

8

VALUE VERSUS VALUES

In a world where data is often positioned as "the new oil," there has been much discussion about the value of data. Data is seen as a source of inferences to improve decision-making and an essential resource for targeted marketing, deep learning, and artificial intelligence. Whole industries have emerged around getting us to share our data. These include digital services and entertainment companies that give us access to free materials in exchange for our personal data, platform economy companies that gather data through transactions, and also business service companies that help traditional sectors adopt platform-based approaches.

In her work on datafication, van Dijck (2014) explains that this technological and institutional apparatus provides the means to gather data, but it also depends on narratives that normalize these practices. In fact, the phrase "data is the new oil" is one of these narratives. Kitchin calls them "data imaginaries": how data is "understood and normatively conceived of within a population or by stakeholders" (2022, 29). These imaginaries focus on the potential for data-sharing to create technological advances and attract investors searching for new frontiers of financial gain. They also orient governments towards enabling regulatory frameworks. For example, data protection laws are designed to strike a balance between the need for companies to collect personal data and people's right to informed consent. This balance is meant to protect individuals while also enabling data collection so that the value of data can be materialized.

All of this talk about the *value* of data distracts us from thinking about the *values* that inform data extraction and the digital economy. Given their focus on profit and growth, companies prioritize values such as efficiency, rationality, freedom, and individualism. This does not mean that companies cannot also value things like kindness, respect, cooperation, or community, but these aren't typically the main values informing corporate planning and decision-making. It's important to recognize the dominant values that inform corporate processes of data extraction because they influence the design of socio-technical systems, as well as our experience of communities of convenience.

THE TECHNIQUE: VALUE VERSUS VALUES

This activity is designed to help us become more conscious about the values underlying socio-technical system design. It also helps us recognize any gaps that might exist between those systems and our own personal values. Underrepresented values can form the basis for new collective efforts, and talking about our personal values can help us form connections and enact communality. In addition, we can engage in value-informed actions, demand that socio-technical systems reorganize themselves around values of local importance or mobilize local values as a basis for policy advocacy.

PROCESS

There are three moments in this activity:
First moment: Understanding the difference between value and values

- Start by taking up a very familiar social arrangement such as the family, a daycare, having a roommate, or being in a classroom. Or select an example that is familiar in your local context. Spend some time talking about the difference between the value of this arrangement and the values that you feel should inform this arrangement. For example, the *value* of having a roommate might be that it makes rent more affordable. But the *values* shaping a co-living arrangement might be trust, respect, kindness, or similar.

- Once the group is familiar with the difference between value and values, turn your attention to the system that brings this group together. Have an initial discussion about value and values in relation to this group. What is the *value* of belonging to your community of convenience? What *values* inform this arrangement?

Second moment: Identify and reflect on dominant versus alternative values in your community of convenience

- Gather examples that describe your community of convenience from advertising, web pages, corporate documents, or the like. Consider some of the common messages and themes that appear in this messaging. How does this enterprise position itself? Create a list of the top 3–5 ideas that you consider central to the messaging of this space.
- Spend some time reflecting on personal experiences in this community of convenience. Does the promotional language match the reality of how this initiative operates?
- Given what you've learned, what are the values that inform this community of convenience? Why do you think that *these* values are central to the organization of this system?

Third moment: Conducting a values-based audit

Share the list of values in exercise 8 with participants. These are a starting point to generate discussion. Have participants identify values that they feel *should* be important to the operation of the community of convenience. Do a round of sharing, and have each person talk about what they think these values should be prioritized in the community of convenience.

- With the entire group, reflect on the implications of the previous steps, discussing:
- Does the operation of this community of convenience align with the values that the group feels are important? How so, or why not?
- How would this system or arrangement need to change to come into alignment with the values of the group?

Exercise 8 A list of common values

Ability	Entertainment	Pride
Acceptability	Enthusiasm	Passivity
Adaptability	Equality	Perfection
Agility	Fairness	Perseverance
Aggressiveness	Excellence	Pessimism
Altruism	Experience	Piety
Appreciation	Extroversion	Pragmatism
Assertiveness	Family	Privacy
Awareness	Faith	Proactivity
Balance	Freedom	Productivity
Benevolence	Generosity	Professionalism
Boldness	Gratitude	Prudence
Bravery	Happiness	Purity
Charity	Harmony	Quality
Compassion	Heroism	Rationality
Competence	Honesty	Rapidity
Complexity	Humility	Realism
Commitment	Imagination	Reciprocity
Connection	Independence	Resilience
Conformity	Individuality	Respect
Continuity	Influence	Responsibility
Control	Inspiration	Righteousness
Conviction	Integrity	Rigour
Cooperation	Intelligence	Security
Courage	Intensity	Simplicity
Creativity	Introversion	Sincerity
Credibility	Intuition	Spirituality
Curiosity	Invention	Solidarity
Dependence	Justice	Spontaneity
Determination	Kindness	Stability
Difference	Loyalty	Strength
Dignity	Leadership	Trust
Diligence	Longevity	Truth
Discipline	Maturity	Vehemence
Discretion	Modesty	Vigour
Disposition	Motivation	Vision
Dominance	Nature	Vitality
Effectiveness	Nobility	Will
Efficiency	Optimism	Youth
Elegance	Overcoming	Others: __________
Empathy	Originality	

- If this system were organized according to the group's values, what would it look like? How would it differ?

Example 8 Value of the InkaFarma Loyalty Program

This example builds on the case study developed by Maria José Lira Moran and Daniela Salas Zuñiga from HiperDerecho in Peru (see section 2). Imagine a feminist collective was exploring datafication of medical services for women. They might initially say that the value of belonging to the loyalty program of InkaFarma pharmacy is access to contraceptives, but that the dominant values informing this arrangement are family planning or consumerism.

The content of the InkaFarma website (https://inkafarma.pe) reveals three observations. First, the website is an online shopping portal that features promotional information about consumer products. Second, InkaFarma offers a service called Aliviamed which conducts medical consults by phone so that you can get a prescription. Third, the website is strongly oriented to women, particularly mothers. For example, it says "In Inkafarma.pe you will find the products that you and your family need" and goes on to promote its "Mom and Baby" category. The main page is dominated by beauty products, diapers, and baby formula.

Based on this, a feminist group might determine that the values informing InkaFarma's promotional materials and loyalty program are consumerism, heteronormativity, and family planning. They might also note that the Aliviamed program enables corporeal autonomy by allowing women to acquire a contraceptive prescription over the phone. If corporeal autonomy and sexual freedom are important values for the group, they may feel that the loyalty program should be more oriented towards sexual health products such as condoms and lubricants. Furthermore, the pharmacy could provide information about sexual health to its users.

COLLECTIVE REFLECTION

Spend some time reflecting on the gap between personal values and the values that inform the operations of your community of convenience, examining what you want to do as a group to address this situation:

- What changes would you like to see in the design or organization of this community of convenience?
- What value-informed activities could you engage in to bring about the changes you seek?

9
OVERFLOW

Socio-technical systems organize us into complex communities of convenience, but these groupings are largely invisible. In our daily lives, what we experience are mundane moments of data-sharing. Though they seem inconsequential, something is expected of us at these junctures. A clerk might ask a simple question such as "What is your ID number?" The immediate expectation is that we will transact our personal data in order to complete the purchase, but larger expectations are also at play, such as consumer loyalty. In that moment of data-sharing, a simple transaction becomes an affiliation, and this affiliation implies certain behaviours and comportments. When we comply, we perform the mandates of a socio-technical machine.

We are often too busy, distracted, relaxed, or otherwise engaged to worry about these moments. However, from time to time, we may experience a heightened awareness of these socio-technical expectations. For example, in Canadian clothing stores, shoppers are often asked to share a phone number or email address, but this is merely a request. It is possible to say no to the clerk, and the purchase can usually be completed without sharing any personal information. But sometimes it is not possible to refuse your personal information and still make a purchase. For instance, if the merchant has adopted a digital payment terminal such as Square, completing the transaction and receiving a receipt may be impossible without providing an email address. This is troubling because, as Square explains in its promotional materials for merchants, "Email addresses and names shared by customers save in your Customer Directory and can be used with Square Marketing or exported for use with a third-party marketing service" (Square Support, n.d.). If you are used to saying "no" to data-sharing and suddenly find that you cannot complete a transaction without sharing personal data, this can be an important moment of awareness.

Everyday moments of data-sharing offer interesting opportunities for unhomed data subjects to reflect on and develop consciousness about the expectations of the various communities of convenience we interact with. Overflow is an activity designed to help us become conscious of how these expectations shape our various experiences of subjectivity. It also helps groups see how socio-technical systems organize their comportment in a datafied society in ways that form them into specific communities of convenience. Giving careful thought to these moments offers us a way to acknowledge data collectives and create the potential for social change.

THE TECHNIQUE: THE OVERFLOW

This activity builds on the concept of "creative spill over" developed by Tomás Villasante (1999). His research sought to identify everyday strategies for creating social change in the context of network societies. Villasante observed that identities and preferences are formed through our social networks and that daily experiences pressure us to conform in the face of social expectations. The power of social networks can make it seem as if identities and preferences are given and unchangeable. However, Villasante argued that citizens "leave open various paths and networks for the construction of social transformation" (1999, 37) in their everyday lives. For example, we can defy expectations by refusing to share personal information during a transaction. Alternatively, we can adopt a subversive position by standing outside of informational capitalism altogether.

For Villasante, acceptance, refusal, and subversion were recognized responses to social pressures, but they created significant barriers to social change. Refusal to conform to expectations puts us in a position of conflict, while subversion leaves us outside society and its benefits. To overcome these challenges, Villasante developed a fourth option called "reversion," a reflective posture that explores how our needs or desires "overflow" the limits of social expectations. Rather than fighting with or avoiding social expectations, reversion suggests that we imagine, enact, or demand changes that accommodate our needs. He states: "What we can build is what social networks are willing to assume and develop from their most deeply felt problems, from their contradictions and needs, thus opening up a plurality of alternatives and diverse processes that will be defined, trimmed or improved, as they are built" (1999, 42).

Overflow and reversion offer important conceptual tools for reflecting on data collection in our daily lives. If we want to apply overflow and reversion in the context of the increasing pervasiveness of socio-technical systems, then we need to be aware of how information systems mediate, invisibilize, and fracture individual and collective processes of identity formation. We experience new forms of social manipulation heavily marked by technological mediation. In this context, a clerk in a store is not part of our social network as conceived by Villasante, but rather functions as part of a socio-technical machine that unites technology and social processes to organize patterns of consumption. We could refuse to share data with a clerk in a store, as suggested by the work of Mann et al. (2003), but this will have little effect. If we want to change socio-technical systems, then what is required is creative participation in socio-technical systems alongside the people we are affiliated with via algorithmic sorting. Thus, recognizing our postmodern data subjectivity and the formation of data collectives becomes the new terrain of defense for personal existence.

Overflow works towards these ends by mobilizing Villasante's categories to help individuals and groups separate what is expected of them by a socio-technical system from what they need or want. In doing so, we are already appropriating technology because we position it as an aspect of a social system rather than a distant and mysterious black box. When this activity is done by the members of a community of convenience, they can begin to identify the potential and limitations of a socio-technical system given their needs or desires. This creates the grounds for creative thinking about the basis for our communities of convenience, people's experiences as members of a data collective, and potential means to achieve social change.

PROCESS

Divide into groups of 5–6 people, and share examples of when you provided data to a private company or digital platform. For example, you might talk about a time when you were asked to state your gender, provide an image, or walk under a surveillance camera. Follow the next steps while reflecting on that particular experience.

- Enter one of these moments into the top left square in exercise 9. This step can be done individually or as a group.
- Now, fill in the remaining white squares by responding to what was expected of you, what you thought or felt at that moment, what you wanted to do, and what you actually did.
- Once you have answered the questions in the white squares, fill in the gray squares with a word that captures your identity at each "junction." For example, if you were expected to comply (What was expected of you?) and your behaviour was compliant (How did you express yourself outwardly?), a possible answer in the second square of the first row might be "compliant."

Exercise 9 Worksheet to identify thoughts and feelings that overflow the system

Data interaction moment:	How did you feel? What were you thinking about?	How did you express yourself? How did you behave?
What was expected of you?		
What did you want to do?		

Example 9 Overflowing thoughts and emotions

Data interaction moment: *Sharing personal data on Facebook when I first signed up.*	**How did you feel? What were you thinking about?** *Skeptical – how would this affect me in the future? What about jobs or school?*	**How did you express yourself? How did you behave?** *I wanted an account, so I provided all the info they asked for!*
What was expected of you? *Share my name, date of birth, image, and thoughts.*	*I felt obligated – like I had to give something up to partake.*	*I just went along with it!*
What did you want to do? *Get an account.*	*Expectant. Excited. I wanted to see what all the fuss was about and connect with my friends online.*	*I wish they provided a space to do a test run. Why do I have to have an account to connect with my friends?*

COLLECTIVE REFLECTION

This activity will help individuals and groups identify the aspects of socio-technical systems that do not sit well with them. By figuring out what we value or need when it comes to socio-technical systems, groups begin to identify alternative futures as well as potential pathways to achieve them. This might include social or technical acts designed to reorient socio-technical systems, but it could also include making demands, both as individuals and as groups. With this in mind, wrap up the activity with the following questions:

- The fourth row of exercise 9 captures what socio-technical systems expect of us, while the fifth row captures our expectations of that system. Does your analysis identify any unmet needs?
- Was there a gap between your actual comportment and what you wanted to do?
- How did the system manipulate your actions based on your feelings, desires, or worries? How does the socio-technical system work to discipline your comportment? Consider both technological and social processes when you answer this question.
- As a data collective, what could be done to change your participation? How would these changes help you resolve your unmet needs and desires?

CONCLUSIONS

Decentring Western Universalism in Critical Data Literacy

In this volume we have mobilized critical design perspectives to study critical data literacy. We did this work because we wanted to locate and develop alternative approaches to critical data literacy that reflect Latin American and Global South experiences, theories, and perspectives. A critical design approach allowed us to scrutinize the assumptions and values that organize our ideas and actions as data subjects, as people doing data literacy, and as researchers, and to consider how these assumptions influence knowing, being, and doing. By combining this analysis with insights from Latin American scholars, we developed alternative starting points for data literacy interventions (the concepts presented in section 2), and designed tools and techniques that manifest data literacy in new ways (section 3).

To conclude this volume, we argue that this work offers a much-needed response to calls to decentre Western universalism in critical data studies. In particular, we address Milan and Treré's (2019) observation that, despite an explosion in critical literature on datafication, these analyses "often take as a frame of reference the liberal democracies of the West, with their sociocultural substrate and long tradition of representative institutions, rule of law, and citizen involvement in public affairs – and their self-representation as wealthy, advanced democracies. The different, the underprivileged, the silenced, the subaltern, and the 'have nots,' whose presence outdoes geographical boundaries, often remain in the blind spot" (320). In response to this call, we argue that critical data literacy has tended to reflect Continental theories of datafication, which focus on surveillance as a challenge to privacy. While surveillance studies offers useful frameworks and concepts for thinking about datafication in any context, it reflects a particular articulation of datafication's problem space, and this has in turn shaped dominant models of community intervention, in this case in the area of data literacy.

In contrast, we argue that *Communing Data Literacy* is an expression of decolonial approaches to datafication. For decolonial scholars, the problem space of datafication is expressed in terms of extractivism, and this opens alternative avenues for thinking about community intervention and change-making. In addition to suggesting different approaches to data literacy interventions, an extractivism frame mobilizes continuities between contemporary forms of datafication and historical expressions of power and control, while also advancing new theorizations of the relationship between technological, cultural, and economic domination. We conclude that, while these theories and concepts were developed with reference to Global South theories and experiences, they offer a welcome set of alternatives for anyone who wants to counter the power of data over our experiences, communities, and relationships.

SURVEILLANCE AND PRIVACY AS UNIVERSALIZING FRAMES

Surveillance and privacy are the dominant frames of reference for critical theories of data literacy coming out of Western scholarship. The tension between surveillance and privacy has long been explored in Continental and North American legal, social justice, and academic work (Lyon 2007). However, beginning in the 1990s observers in the fields of market research, political economy of technology, and digital rights began to argue that data mining driven by novel applications of information technology would create new challenges to personal privacy (Gandy 1993). These fears were justified in 2018 by the Cambridge Analytica scandal, when former Cambridge Analytica employee Christopher Wylie revealed that the company was using Facebook's social media platform and its Open Graph program to build psychological profiles of tens of millions of users.

In response, the then-emerging field of surveillance studies argued that while surveillance has always happened, we are experiencing fundamental changes to how it manifests. In particular, given the rise of big data, the internet of things, and artificial intelligence, "surveillance capitalism" (Zuboff 2019) and its underlying structures of "dataveillance" (van Dijck 2014) are said to be ubiquitous. Indeed, Couldry and Mejias refer to datafication as the "God view" of social interaction that, "though parceled out among many corporations, coalesces in a corporate social imaginary that justifies new ways of governing everyday life for the benefit and convenience of corporate power" (2019, 190).

The word "surveillance" is ideologically charged. It calls to mind forms of control that take away personal freedoms. Given this starting point, at their root, Continental theories of datafication address enduring questions such as "how are we governed?" and by extension, "to what extent are we free?" In these works, Foucault, Deleuze, and Baudrillard are taken up to explain how symbols and information systems create or perpetuate social orders and forms of domination in what otherwise appear to be free societies. Indeed, this literature features a tension between Foucauldian thought, which is used to explain subjugation through algorithmic "social sorting" (Lyon 2002), and Deleuze's vision of a new society of control enacted through "data-doubles" (Poster 1990), "simulated surveillance" (Bogard 1996), and consumer seduction (Boyne 2000; Han 2017). These works argue that our perception of freedom is mediated by mechanisms that order our social experience. Foucault argued that imagery and architectural forms used representations of real forces to compel obedience, so, for example, depictions of violence or architectural forms suggesting surveillance were used to compel orderly conduct. But for Baudrillard (1994), in contemporary society those indications are mobilized through mental images that have no clear link to material realities, and thus create an experience that *simulates* a sense of surveillance rather than depicting actual surveillance.

While the sources or mechanisms of unfreedom can be debated, the point here is that *the problem space for surveillance studies is fundamentally tied to the question of how freedoms are governed.* Accordingly, surveillance studies argues that ubiquitous monitoring normalizes generalized discipline or generalized apathy, and that these modes of control have negative implications for freedom and social justice. And, since datafication lies at the heart of surveillance capitalism – since it is the mechanism through which surveillance is articulated – surveillance studies positions it as a major contemporary source of control and unfreedom.

This characteristic of surveillance studies has had important implications for how datafication came to be discussed, debated, and regulated in North America and Europe. The use and collection of personal data by governments and corporations in European and North American democracies has been discussed and regulated since at least the 1970s. However, after the 2001 dot-com stock market crash, the internet was redefined as Web 2.0, and monetization of data was identified as a new source of economic growth for digital enterprises (O'Reilly 2005). As these new forms of capitalization unfolded, observers from diverse fields such as consumer research (Turow 2003; Turow and Hennessy 2007; Turow, Feldman, and Meltzer 2005), youth studies (Grimes and Shade 2005; Chung and Grimes 2006; Latour 2007), and gender studies (Southworth et al. 2005) began to identify significant new risks to vulnerable populations from fraud, undue influence, and digital violence. This gave rise to policy debates in North America and Europe that were heavily focused on the protection of individual rights. Since policy debates were framed in terms of individual rights, and academic debates were largely framed in terms of surveillance, the problem space for datafication came to be structured as a tension between privacy and surveillance, such that the protection of the former was seen as the solution to the latter. According to this logic, if privacy is protected, then surveillance is resolved.

The story does not end there, however. Given the importance of data as an economic resource in the post–dot-com digital economy, governments found themselves in the position of arbitrating between the individual right to privacy and corporate imperatives to drive innovation and economic productivity (Mansell and Steinmueller 2020). As Hare argued in a 2019 editorial in *The Guardian*, the European Union's General Data Protection Regulation (GDPR) could have made "privacy the default and requir[ed] us to opt in if we want to have our data collected. But this would hurt the ability of governments and companies to know about us and predict and manipulate our behaviour" (Hare 2019). As a result, personal data protection laws have come to enable data collection as long as certain minimal conditions are met. Of particular note, the European Union's 2018 General Data Protection Regulation offered a model for personal data protection based on the consent-based ARCO framework:

third parties must obtain an individual's consent to collect their personal data, and individuals have the right to access the personal data that an entity holds about them, rectify any errors in that data, have their data removed from the database, or oppose the use of their data for certain purposes. In total, where surveillance studies theorized datafication in terms of its potential to create unfreedoms, policy-makers sought to protect freedoms by regulating corporate access to personal data.

In these legal frameworks, the onus for personal data protection rests largely with the individual, who is left to figure out how best to mobilize their new digital rights. In other words, they have been built on the assumption of a coherent, individual data subject who is a rights-holder and rights-seeker. Thus, "data subjects" must make independent decisions about data-sharing in each of the many digitized transactions that they encounter on a daily basis. As we've discussed elsewhere in this volume, this clinical view of a "data subject" reduces a person to a source of data, which hides from view the power imbalance created between distant and opaque databases that organize systems of transactions, and the emotively informed but atomized individuals who must find ways to respond to myriad digital requests. It also hides from view the different ways that people's "data bodies" are interpolated by these databases, and how this re/produces reductive logics, intersectional inequalities, and historical harms (Lewis et al. 2018).

Of particular concern to surveillance studies, even when these power imbalances are made plain, people tend to shrug their shoulders at the prospect of being surveilled through their digitally mediated activities. There is also a "privacy paradox" between people's stated desire for privacy and their tendency to pursue online activities that are exposing and therefore risky (Barnes 2006). Some authors suggest that these tendencies represent a pragmatic response by data subjects who mobilize their personal data as a resource that can be traded strategically in daily negotiations (H. Taylor 2003). They may, for example, feel that the costs of self-protection and the benefits of data-sharing outweigh the possible risks. Others argue that this is part of a larger pattern of surveillance realism in which people have become resigned to a newly legitimated hegemony of surveillance capitalism (Mundie 2014; Dencik 2018) or acculturated into the new daily acts of data-sharing that have become a part of digital transactions. These forms of legitimation and acculturation then become socio-technical stabilizations of patterns of control and domination that have systemic implications for human rights and social justice.

As a result, surveillance studies is deeply invested in the question of how to respond to the new forms of control and domination – or unfreedoms – enabled by the collection of personal data. There have been many nods to "everyday forms of resistance" in the literature (Martin, Brakel, and Bernhard 2009). In this vein, the idea of sousveillance (drawing on the French *sous* for "below") has been put forward as a means to "look back at" or "look up at" the panoptic eye of data-driven surveillance (Mann, Nolan, and Wellman 2003). The goal of the artistic work that gave rise to the term "sousveillance" was to mirror patterns of control back at bureaucracies, and in this way provoke reflection about how technology is used to organize and discipline. For example, in one experiment, Mann walked around with a video camera on his body to draw attention to growing video surveillance and create public dialogue. The notion here is that hegemonic oversight can be countered by rhizomatic expressions of resistance. However, while it is true that people do often say "no" to having their data collected, these disparate instances fall short of widespread expressions of agency or literacy in response to datafication. What is more, because these responses rely on the individual agency of data-holders, they play into the very logic of consent-driven individual data collection.

Lyon (2007) suggests that rather than reflecting a balance of power between data collectors and data subjects, the tension between sousveillance and surveillance points at the emergence of a generalized condition of societal surveillance, or what others have called "liquid surveillance" (Bauman and Lyon 2012) or the "audit society" (Power 1999). In a society of digitally enabled and artificially intelligent voyeurs, we are continually scrutinized and categorized, and this has made us all more watchful, critical, and judgmental. In such a context, the metaphors of above and below or broadcaster and audience become strained, because the systems of control are pervasive. Of notable concern, to the extent that these systems of control are driven by simulated surveillance (Bogard 1996), then our individualized watchfulness will be grounded in a veritable house of mirrors of indeterminate or "simulated" criteria, and thus rather than our being united around the values, needs, or "truths" that inform our views of the world, watchfulness becomes a source of divisiveness, isolation, or apathy, contributing to a generalized crisis of disinformation, polarization, and alienation that poses a fundamental threat to democratic governance.

With this in mind, surveillance studies explores the changing nature of democracy and the public sphere through work on data justice (Dencik 2019) and data activism (Milan and Gutiérrez 2015; Milan 2019; Milan and van der Velden 2016). As the members of the Data Justice Lab at Cardiff University explain, "a focus on data justice advances an orientation towards destabilizing the appearance of natural orders, sets out to reveal contemporary power relations, and considers how they could be different" (Dencik et al. 2022). These works explore how people can become more aware of and more involved in asserting their values, needs, and desires with regards to the new patterns of order and domination that are shaping their experience of freedom. In total, then, the universal frame for critical data studies prioritizes figuring out how to rethink governance in ways that ensure, protect, and promote freedoms. We argue that this focus has also strongly influenced the emerging field of critical data literacy studies.

CRITICAL DATA LITERACY AS A SEARCH FOR FREEDOM

In their work on data colonialism, Couldry and Mejias observe that "[t]here is an emerging industry dedicated to advising us how to better deal with social quantification" (2019, 194). They aren't wrong. Prior to 2018, the concept of personal data literacy did not exist in academic literature, but a robust sub-field has rapidly emerged to fill this void. It explores the challenge of addressing the unfreedoms created by datafication, either by helping people acquire skills that protect their privacy or by building critical insights that allow people to express agency and participate more readily in citizenship and democracy.

Literacy works focused on skills acquisition argue that individuals require new knowledge to navigate the digital transformations enabled by datafication. This includes information literacy to locate, acquire, and evaluate data; contextual literacy to make sense of social processes such as privacy and surveillance; and media literacy to make sense of messaging around data use. These skills are what are known as functional literacies, meaning that we require them to function in a particular context (Johnson 2007). Personal data literacy initially emerged as an extension of these concerns, and it focused specifically on people's ability to manage the increasing use of their personal data in digital transactions (Bartsch and Dienlin 2016).

Work in this vein has been inspired by Park (2013), who showed that users with greater knowledge of digital systems were more likely to protect their personal data during online transactions. Drawing on Park's work, Trepte et al. defined online privacy literacy as the competencies necessary to close knowledge gaps relevant to the digital economy: a "combination of factual or declarative ('knowing that') and procedural ('knowing how') knowledge about online privacy" (2015, 339). Accordingly, online personal data literacy requires the following competencies: knowledge of institutional practices, the technical aspects of online privacy and data protection, legal frameworks, and strategies to manage individual privacy. This work formed the basis for the Online Privacy Literacy Scale (OPLIS), which assesses knowledge about institutional practices, the technical aspects of data protection, data protection law, and data protection strategies.

These functionalist frameworks are critiqued on multiple grounds, such as being overly liberal, focusing too much on risk, and being uncritical. Masur (2020) argues that "contemporary discussion of online privacy almost exclusively adopts a perspective that is rooted in liberal theories" (258) and thus it defines privacy as a negative freedom in which "the focus is protection against access to and identification of the individual" (259). Citing Park and Trepte et al., he concludes that this leads to the proposed solution, which is limited to "strengthening individuals' knowledge, skills and abilities to protect themselves" (Masur 2020, 259). Similarly, Pangrazio and Selwyn argue that "there is clearly a need to better support individuals to engage critically with their personal data, so they have a sense of understanding, control and agency with the data assemblage," and as such, "data literacies need to include critical understandings of the reconstitutions and recirculation of data," which include "conceptualisations of the inherently political nature of the broader data assemblage" (2018, 8). In other words, rather than simply teaching people how to protect themselves, data literacy should build up people's ability to make critically informed decisions about their personal data.

Accordingly, critical data literacy has emphasized building capacity for self-directed learning, the ability to engage with new cultural forms and social practices, and critical thinking skills. As Martin argues, when data literacy moves past functionality into criticality, it emerges "as a powerful tool for the individual and the group to understand their own relationship to the digital: to be aware of the role of the digital in their own development and to control it; to place the digital at the disposal of their own goals and visions" (2008, 156). In this vein, Pangrazio and Selwyn draw attention to the need for greater focus on data reflexivity, data strategies, and data tactics that enhance the legibility of digital systems and enable people to respond to them (2019; 2023). Data reflexivity involves considering the implications of the circulation of different types of personal data through different systems for oneself and others, while data strategies and tactics involve working purposefully with data, including one's own, to achieve personal and social objectives. Masur similarly highlights the need for critical privacy literacy focused on the ability to "identify, question, and criticize societal structures, processes, norms and practices that affect individuals [*sic*] privacy on a societal level" as well as to express "social responsibility and motivation to take become [*sic*] agents of social change" (2020, 261).

In both its functionalist and critical guises, the literature on personal data literacy that has emerged in the context of anxieties about surveillance and privacy has focused on individuals, activists, or citizens as both people at risk and people who need to equip themselves with new skill sets to live in a surveillance society. They have prioritized cognitive abilities and critical thinking as an avenue to overcome the unfreedoms created by the digital mediation of social processes, and they have looked to individual strategic acts, expressions of citizenship, everyday tactics, and creative expression as ways to counter datafication's impact on justice and freedom.

These approaches haven't been well-received in some quarters. For one thing, they are seen as a way of accommodating people to datafication, which is itself treated as a form of digital transformation. As Couldry and Mejias (2019) put it: "we should not pretend that media literacy is a means to resist data colonialism; at most it enables us to live more at ease with it, thus normalizing it for the longer term" (194). For example, Pangrazio and Selwyn define critical data literacies as "being able to unpack the politics of everyday texts and, most important, for these understandings and disposition to inform people's digital practices" (2023, 84). This positions digital literacy as a strategy to make sense of changes that are already taking place and prevent the worst of their potential harms while giving an individual some sense of agency over their relationship with the changes.

Meanwhile, in a talk at the South by Southwest Education Conference, danah boyd pointed out that, "[n]o matter what worldview or way of knowing someone holds dear, they always believe that they are engaging in critical thinking when developing a sense of what is right and wrong, true and false, honest and deceptive. But much of what they conclude may be more rooted in their way of knowing than any specific source of information. If we're not careful, 'media literacy' and 'critical thinking' will simply be deployed as an assertion of authority over epistemology" (2018, n.p.). Similarly, Couldry and Mejias argue that "media literacy relies on the virtuous 'disposition' of the subject, which misses how the new order works to dismantle the autonomy of the subject" (2019, 194). In this sense, personal data literacy overlooks how new forms of digital mediation reshape the potential for individuals to express their agency. We would add that data literacy scholarship tends to depart from an impoverished ontology of datafication which treats data like a mechanism or resource that flows through tools that are separate from humans. As a result, data literacy has tended to fall short of responding to the new forms of subjectivity and socio-technical reality manifested through people's (often unavoidable) co-constitution of datafied systems.

Rather than reject all forms of data literacy, however, we feel that it should be rethought in ways that respond to the reality of these "dismantled subjects" – the unhomed data subjects we discussed earlier on. As boyd said in the conclusion to her talk at SXSW-Edu in 2017, "the key is to realize the humanity in ourselves and in others. We cannot and should not assert authority over epistemology, but we can encourage our students to be more aware of how interpretation is socially constructed" (2018, n.p.) To do that, we feel that we can benefit greatly from alternative starting points that move away from the privacy and surveillance frame that shapes much of the European and North American work on data literacy. But before we get to that, it is important to consider why the universal frame has dominated discussions about critical data literacy, and the challenges of applying this frame without careful consideration of context.

QUESTIONING THE UNIVERSAL FRAME

We argue that extractivism and autonomy offer robust alternatives to surveillance and privacy frames as starting points for research and engagement related to datafication. We will explore these alternative frames further along, but before turning to that discussion, this section first considers why Western and liberal ideas have been so important to critical data studies. We also explore some of the reasons why these ideas fall short in many Latin American contexts.

The focus on surveillance and privacy in critical data studies happens in large part because the digital innovations and capitalist impulses that sparked our current experience of datafication arose primarily in North America and Europe. Mirroring the history of the spread of the physical internet (Press 1997), their rollout started in these regions and from there diffused to the rest of the world. As has been amply observed (Ricaurte 2022; Taylor and Broeders 2015), the diffusion of these technologies carries with it the ontological assumptions that go into the development of the machines, which has implications for the organization of the knowledge capital required to appropriate them. Not only that, but these processes also come accompanied by a series of narratives and counternarratives that circulate internationally between observers, academics, advocates, and officials through global meeting spaces such as the Internet Governance Forum (IGF) or RightsCon. Thus, problematizations and theorizations of datafication were carried into Latin America over the same roads that carried the datafication itself. For example, when the European Union adopted the GDPR in response to Continental concerns about personal data privacy, governments throughout Latin America followed suit, using the same narratives to justify the implementation of personal data protection laws (Carrillo and Jackson 2022), many of which copy foreign legal language word for word.

Much of critical data studies similarly draws on Western theories and contexts (Milan and Treré 2019), and positions datafication as something that has already happened in particular ways. As van Dijck says in her seminal work on datafication, people's tolerance for datafication and their willingness to share their personal data with public officials and corporate actors result from "the gradual normalization of datafication as a new paradigm in science and society" (2014, 198). She goes on to position datafication as a new institutional paradigm that "is becoming a leading principle, not just amongst techno-adepts, but also amongst scholars who see datafication as a revolutionary research opportunity to investigate human conduct" (198).

But which society is she referring to when she shares these observations? Is this a global paradigm? If datafication presents a new set of global principles or opportunities, then digital transformation (Vial 2019) must be doing the work of reorganizing local institutions, processes, strategies, and goals around those principles and opportunities. Is that happening? Are processes of digital transformation actually unfolding in local contexts, and if so, how do people actually experience them? These questions call for a critical reconsideration of the intellectual orientation provided by critical data studies in general and surveillance studies in particular.

Because we realized this project in Latin America, here we consider the central assumptions about datafication with reference to the political economy of that region. We find that key assumptions about datafication are difficult to sustain given specific patterns of digital uptake in Latin America, the weak ideological force of the surveillance narrative in that context, and the experience of rights-bearing subjects in the region, each of which are explored on continuation.

First, smartphones have extended social media channels and platform services to wide swaths of Latin America's population, and datafication is also influencing banking and the delivery of social services such as healthcare. Depending on the extent and nature of digital transformations, the effects of datafication approach the European experience in some Latin American countries such as Uruguay (Rivoir and Landinelli 2017). However, in other contexts the experience is quite different. For example, in Peru's corporate context, retail sectors are cautious about investing in predictive technologies (Martin Valdez 2020; UTEC and Everis 2021). This is likely because predictive marketing isn't necessary for companies that already have oligarchic control over the distribution of goods and services, and what is more, when economies are unstable, capital will avoid the risks associated with large-scale technological investments. Instead, domestic industries are more likely to take up loyalty programs or implement business efficiencies in order to defend oligarchic arrangements. This means that in much of their daily lives, people experience datafication not as a source of surveillance per se, but rather as a form of advertising, and their concerns aren't about privacy so much as they are about fair pricing and good treatment.

Second, it is difficult to sustain the Continental narrative of surveillance in contexts marked by historical experiences with dictatorship, civil conflict, and economic regress, as is the case in many Latin American countries. In these contexts, the potential of digital innovation to improve the delivery of lower-priced goods and services is welcomed with open arms by many sectors of the population. For example, for Peru's newly emergent middle class, which came into being in the context of economic growth under neoliberalism, moving past the country's protracted period of civil conflict and economic decline has meant embracing market-oriented reforms (Carranza Ko 2013; Lust 2019; Roberts 1995). Rather than asking whether datafication *should* take place, the country's middle class is taken up with debates about how best to ensure that digital transformation happens, and that it supports continued economic transformation.

Meanwhile, human rights activists and members of Peru's Indigenous communities argue that it is not possible to reconcile and move forward until the injustices of the civil conflict are fully recognized. Despite this, Laurie and Bonnett (2002) argue that it is difficult for Peruvian Indigenous and social justice activists to confront the power of discourses that promise concrete advantages in the form of better access to goods and services. As they say, "when today's generation of activists say, 'we don't have a history,' they are reflecting, not the reality of Peru's past, but the ideologies of and influences upon its present" (49). In this context, if data-sharing secures a discount, then so much the better. For many, the kinds of "controls" asserted by datafication are hardly anything to complain about, especially in comparison to the excesses of the authoritarian period in the region. And anyway, continued challenges with the uptake of digital technologies make people doubt the ability of domestic governments or domestic corporations to mobilize datafication effectively, so the risks of surveillance are considered to be low. There is a general sense that intelligence collected by transnational actors is likely to be incomplete, ineffective, or unused in the region. Indeed, it is sometimes suggested that the global majority's position on the global periphery positions them beyond the purview of the digital metropole, for better or for worse.

Finally, the notion of an autonomous rights-bearing subject also demands scrutiny in the context of debates about datafication in Latin America. In the context of tenuous democracies and widespread and blatant social and economic injustices, human rights form an essential and constant touchstone for advocacy agendas and public discourse in the region. However, it is also understood that legal provisions are never sufficient as grounds for advancing human rights, and that collective rights are as important as individual rights, or more so. Latin America's colonial and post-colonial experiences have entrenched forms of social and economic inequality, racism, and misogyny that place large groups of people beyond attaining full access to citizenship rights. Indeed, the very ideal of citizenship is challenged by colonial histories, processes of state formation after independence, contemporary patterns of migration, and widespread failures of governance. Without larger processes of economic and social transformation, legal instruments are a necessary but insufficient solution to the achievement of all rights, including digital rights, and these legal rights can only be effective to the extent that they reflect collective autonomy.

This of course does not mean that datafication isn't happening in Latin America. Clearly smartphones are widely used in the region, and digital technologies are being taken up in telecommunications, banking, retail, and entertainment. Latin America is home to platform businesses such as Rappi (based in Colombia) and MercadoLibre (based in Argentina) that rival their global corollaries in size and reach. And the region has seen rapid growth in digital services companies that support digitization and datafication. Still, we can't assume that the logics, applications, and effects of datafication will be the same in one context as they are elsewhere in the world. Furthermore, we can't assume that people will understand, experience, and be affected by them in the same ways in Latin America as elsewhere.

In total, the surveillance frame is a hard sell for thinking about datafication in Latin America. And this also means that the idea of unfreedoms is insufficient as a starting point for thinking about critical data literacy. You would have to believe that functional literacies, agency, or citizenship were sufficient to overcome the actual challenges that people face in their everyday lives. But unfortunately, given local expressions of oligarchy, underdevelopment, or injustice – all of which stem from colonial histories – Western approaches to critical data literacy fall well short of the mark for the Latin American context. Building from this analysis, we suspect that they fall well short of the mark for any group which finds that the dominant governing regime, whether cultural or institutional, doesn't serve their need for recognition or redistributive justice (Fraser 1995; 1998). In these instances, what is required is an approach to data literacy that helps people understand datafication's role in (re)producing or (re)organizing social structures, and how those produce or sustain inequitable social relations. This necessitates a very different framing of the problem of datafication, and an alternative starting point for doing data literacy work.

EXTRACTIVISM AND AUTONOMY AS ALTERNATIVE FRAMES

In contrast to the surveillance literature, with its anxieties about privacy and freedom, data extractivism puts the focus on how new systems of subjugation organize people in ways that facilitate digital dispossession (Thatcher, O'Sullivan, and Mahmoudi 2016) and the commodification of life (Couldry and Mejias 2019). Here the focus has shifted from freedom, which is a liberal concern, to autonomy and self-determination, which are material concerns. As Couldry and Mejias put it, "Through data colonialism, contemporary capitalism promises to consume its last remaining 'outside,' dispossessing human subjects of their capacity as independent sites of thought and action" (6).

Once again, the term "extractivism" functions as an ideological marker for the articulation of a problem space. It is not actually possible to "extract" data in the way that, say, iron ore is extracted. When you tell someone your name, they aren't removing your name from you, but rather they are establishing a marker that allows them to differentiate you from another person. However, "data extractivism" calls attention to how algorithmically informed processes of data generation and analysis re/produce relations of inequality in how they organize epistemic channels, generate intelligence, and assert ways of knowing (Ricaurte 2019). These processes reflect and build upon long-standing patterns of extractivism in which technological and epistemic assertion serve to facilitate economic flows and political domination.

With this in mind, data extractivism focuses on the epistemic and material implications of datafication for the data commons and communicative sovereignty (Reilly 2024). Materially, information systems are being organized around the economic priorities of transnationalized interests to the detriment of local communities. The most dramatic example of this comes from the establishment of cloud computing facilities in cities such as Santiago, Chile, given their negative impact on the natural environment, and how they divert key resources such as electricity and water away from local populations (Arellano, Cifuentes, and Ríos 2020; Lehuedé 2022). Epistemologically, the creation of data doubles, social sorting, and the use of simulated surveillance (as identified by Continental theorists and discussed in the previous section) function not so much to *control* domestic societies (although this may happen as well), but rather to organize epistemic processes according to the needs of capital so that global peripheries can continue to subsidize the twinned wealth and intellectual domination of capital. In doing so, these processes displace and devalue local processes of meaning-making and leave people less able to make sense of and appropriate technologies and ways of knowing to serve their own ends. This means that the central problem from a data extractivism point of view isn't to secure people's human right to privacy and authentic experience of freedom, as in the case of a surveillance frame – nor, incidentally, is it to address disinformation, which has long been central to the experience of colonialism, imperialism, and authoritarianism and therefore doesn't constitute a new challenge. Rather, the central problem is to reappropriate the data commons and achieve data sovereignty as a basis for the creation of a generative knowledge commons.

This knowledge commons is not simply a shared "field" of cached knowledge resources that can be drawn on by individuals (as the idea of open data sometimes suggests), but rather it is a space that is cultivated through situated social connection, sharing, dialogue, and collaboration. Data is not treated simply as a material resource – the "new oil" – though this is absolutely part of the discourse. For example, in an editorial for *Al Jazeera*, Mejias (2023) argues that the sharing economy was a ruse to dispossess us of the public data needed to train AI models. As he says, in the digital economy, "[o]ur data has not only been appropriated, but is increasingly used against us," while Big Tech clings to the argument that "data extractivism is all done for our benefit." But more to the point, these material considerations gain their significance by way of Latin America's postcolonial and critical communications literature, in which data is but one element of the epistemic terrain. Drawing on Martín-Barbero (1993), the commons can be idealized as the most democratic means possible to express the sensorium that organizes our human capacity for perception. Different patterns of technologically facilitated appropriation of the commons mediate how we perceive the world, and in this sense, Big Tech supplants the commons in ways that appropriate not just data but also people's capacity to perceive the world creatively and independently. Importantly, in this view, the commons is not a library full of books or a database full of figures, but rather *a space in which perceptive beings commune with each other to manifest human spaces of interaction.*

This then ties to Mignolo's (2005) work on *The Idea of Latin America*, which argues that postcoloniality in that region was set in motion by newly independent elites who emulated European art, architecture, and philosophy, up to and including the "idea" of being *Latin* and part of the *Americas*. In looking to Europe, these mestizo elites turned their backs on their own heritage and closed down spaces of creative engagement with the mestizo and Indigenous majority, whose energy animated the region's cultural potential. Mignolo argues that to move past this colonial rupture, this original sin of European emulation needs to be redressed, a challenge that post-coloniality in Latin America sees as being tied to decolonization of thought and to the restoration or reanimation of ways of being, thinking, and knowing – particularly ones that draw on the pluriverse of other ways of knowing (Escobar 2018) – and which may in fact flow through the very spaces of informality, margination, corruption, and erasure that so clearly mark daily life in the region. These are human processes that must be mobilized or confronted to restore the epistemic heart of the commons through moments of engagement, dialogue, creativity, and care. They in turn offer an antidote to the disarticulation wrought by societal surveillance (Bauman and Lyon 2012), digital dispossession (Thatcher, O'Sullivan, and Mahmoudi 2016), and the capitalization of life (Couldry and Mejias 2019).

While the problem space for datafication is organized very differently in the case of data extractivism, this does not mean that Continental theory has nothing to contribute to this discussion. Where Continental strands of critical data studies express concern about how new patterns of order and domination threaten democracy and the public sphere by generating isolation, division, and apathy (Calhoun 1998), theorists of the data commons similarly worry that datafication establishes the conditions for a "zombie democracy" wherein citizens are so guided by the logics of consumption that they cede autonomy over independent thought and stop engaging in collective cultural production (Rincón 2018; Ricaurte 2022). For example, Couldry documents a 2017 conversation with Martín-Barbero about how digital platforms are reshaping "the very spaces where culture and social life are lived out" (Couldry 2019, 187).

What unites critical scholarship is the concern that datafication and other forms of digital mediation displace our ability to perceive the world in terms of human sympathy and connection, and to individually or collectively negotiate with or express ourselves in the face of centralized sources of power. It is not that digital apparatuses are emotionless (clearly digital spaces are emotionally charged) but rather that their formulation contributes to a sensorium and commons in which it is difficult to encounter spaces of care, connection, sympathy, and expression. So, while surveillance studies yearns for freedom, and extractivism reaches for autonomy, in the end, they share a common understanding of the central challenge of datafication: how to understand and overcome the novel subjectivizing forces of the new digital economy and their tendency to engender disconnection, divisiveness, isolation, or apathy. The difference, however, lies in how critical perspectives address this problem – whether they see it as a challenge of individual unfreedoms or as a lack of self-determination, and whether solutions lie in engendering critical citizenship, or in fostering pluriversal forms of connection that enable a multitude of knowledge commons.

THINKING AND *DOING* DATA LITERACY DIFFERENTLY

Thinking and doing data literacy differently means exploring how datafication mediates our capacity to know and make sense of our human experience, and what this means for the organization of the knowledge commons. Achieving hermeneutic autonomy and epistemic justice requires an understanding of how datafication structures our ways of knowing: how it shapes moments of individual perception, how it occupies our subjectivity, what it makes legible or illegible, how it organizes social and cultural relations, and how it organizes the knowledge economy. If literacy interventions are to offer a critical response to datafication's multiple mediations of the sensorium, then they need to find ways to help people recover their autonomy and come together around the production of grounded data generated from local experiences, emotions, and opinions. And they need to offer people an opportunity to commune that knowledge in ways that create knowing, sensing, and creating communities. In this way, critical data literacy can counter the atomizing effects and extractivist logics of datafication, and its multiple appropriations, enclosures, and exploitations of the knowledge commons, and begin to create the foundations for a plurality of alternative ways of thinking, being, and doing.

In making these arguments, we are channelling, refiguring, and updating discussions that have long circulated throughout Latin America and beyond, and in some instances making them accessible to an English-speaking audience. We have been greatly inspired by the postcolonial teachings and Global South perspectives of several essential authors from Latin America as well as our colleagues in the Tierra Común Data Colonialism network, and our collaborators in digital justice NGOs throughout the region. We are also drawing on the community-based and intellectual works of various actors who are exploring alternative approaches to critical data literacy. The nine techniques we outlined in this volume are just a sampling of many possible ways to commune data literacy, and we hope that they will act more as a jumping-off point for ongoing experimentation than as a how-to guide. We also hope that our experiment in critical design research offers an example of how to engage in this type of work. To conclude, we acknowledge other works that commune data literacy and use them to exemplify the common characteristics of this approach.

We are greatly inspired by arts-based approaches that provoke us to think about how data and technology are changing our social relations (Sander 2020). Research-creation is a fruitful avenue for communing data literacy because it invokes our visceral relationship to the knowledge commons and offers a means to "see" and make legible datafication's many effects. These efforts confront extractive forces by animating the connective commons and enacting creative autonomy (Bhargava and D'Ignazio 2015; Bhargava et al. 2015; Tygel and Kirsch 2015). For example, in one of its data literacy projects, the Connection Lab run by Raul Bhargava explored story-telling and muralism as ways to build data literacy among students at public schools in Belo Horizonte, Brazil. Bhargava emphasized the participatory nature of the project, how it de-centred technology as a means to work with data, and the way it showed students how data can support social change (Bhargava et al. 2016).

Another research-creation example is provided by Mona Hedayati's work on data visceralization (Hedayati 2025). In an artistic performance called *Curves & Reverbs*, Hedayati asked participants to wear biosensors that captured neurophysiological data and transmitted their affective biological responses as sound. She then blended these feeds with her own neurophysiological data to produce a sonic experience. The performance makes individual and group affective data legible to the participants, challenging data's reductive tendencies, and also creating a space to build empathy and understanding between participants.

We are also inspired by works of technological appropriation that respond to a global context in which "technical knowledge of how to work effectively with data is in the hands of a small class of specialists" (D'Ignazio 2017, 6), and little attention is paid to "who has access to the computing power and technical know-how to be able to make sense of data and how they come to acquire and deploy that knowledge" (7). For example, Tironi and Valderrama developed an application that simulates the algorithmic profiling and recommendation systems of large platforms. This allowed them to convene "open spaces for decolonial doing, thinking, and feeling that problematize the colonial principles and logics on which these algorithmic systems are founded" (2021, 6–7). They concluded that, "instead of promoting technical skills or knowledge acquisition, critical data or algorithmic literacy from the South must advance local processes of autonomy and emancipation and enable problematization of the agendas, values and ideologies implied by these algorithmic systems" (27). As in the case of research-creation, these technological projects intervene in ways that bring people together and enable new understanding. Similarly, we appreciate the work of Sursiendo, a group of software activists from Chiapas, Mexico, that pursue digital communality, collective digital rights, and hacker feminism. As they explain, their work "aims to create small nodes of digital autonomy that enable us, step by step, to gain control of our connected lives and rewrite technologies according to our traditions, concerns and needs" (Sursiendo, n.d.).

For us these are all examples of efforts to commune data literacy because they mirror or express the concepts we've explored and mobilized throughout this book. They exhibit a set of principles that we argue reflect a critical commitment to rearticulating knowledge commons and reclaiming the autonomy and creative potential of people and communities. They reflect the key characteristics of communing data literacy, which are as follows:

Communing data literacy avoids treating people like data subjects, and instead engages them as complex thinking beings who are trying, in multitudinous ways, to make sense of their feelings, experiences, and opinions under conditions of datafication. And as such, communing data literacy doesn't set out to tell people what data is, or assume what data might mean to people, or predict how they experience it. Instead, it starts by curating experiences of the transformations that result from datafication and uses these observations as a starting point for thinking about and rethinking social relations.

Communing data literacy asks people to pay attention to how datafication organizes relationships between individuals and the groups they belong to, and between individuals and other members of those groups. It doesn't assume that the members of these groups will know each other or even share much in common, but focuses instead on the logics that convene people, and how those logics affect them. Indeed, it strives to make datafication's effects legible through carefully considered moments of sharing and dialogue. In doing this work, communing data literacy understands that data or technology isn't the point of a data literacy intervention; the goal is to generate a knowing and sensing knowledge commons that is grounded in meaningful social connections. There can be many ways of convening people and making data's effects legible, and indeed there are many people doing this work in a rich variety of ways. There need to be as many ways of doing this work as there are groups of people seeking to understand their experiences. In this sense, this book doesn't aspire to guide people, but rather offers an invitation to start a conversation.

Finally, while participants in these spaces may decide to articulate clear demands about how data is collected, interpreted, and used as a result of communing with one another, we acknowledge that not all efforts to commune data literacy will result in the launch of new initiatives or collective responses. We would love it if they did, and we think that coming together is an important first step in opening the way to collective actions. However, communing data literacy doesn't need to result in projects for social change to happen, because the act of generating alternative sources of understanding and entering into dialogue already counters the extractivist and atomizing logics of datafication. The larger goal of collective engagements with datafication is to ground the knowledge commons in local ways of knowing, being, and doing, because in this way we are better able to achieve material and cultural autonomy and self-determination. Ultimately, sharing everyday experiences and exploring data values is a necessary first step for building data literacy, and will commune the types of data sovereignties, data collectives, and data cooperatives that empower people and enable social change.

REFERENCES

Ahmed, Sara. 2021. *Complaint!* Duke University Press.

Aimé, Isabelle, Fabienne Berger-Remy, and Marie-Eve Laporte. 2022. "The Brand, the Persona and the Algorithm: How Datafication Is Reconfiguring Marketing Work ☆." *Journal of Business Research* 145 (June): 814–27. https://doi.org/10.1016/j.jbusres.2022.03.047.

Akbari, Azadeh. 2020. "Follow the Thing: Data: Contestations over Data from the Global South." *Antipode* 52 (2): 408–29. https://doi.org/10.1111/anti.12596.

Alaimo, Cristina, and Jannis Kallinikos. 2017. "Computing the Everyday: Social Media as Data Platforms." *The Information Society* 33 (4): 175–91. https://doi.org/10.1080/01972243.2017.1318327.

Animikii. 2019. "Decolonizing Digital: Empowering Indigeneity Through Data Sovereignty." https://www.animikii.com/news/decolonizing-digital-empowering-indigeneity-through-data-sovereignty.

Arellano, Alberto, Lucas Cifuentes, and Critóbal Ríos. 2020. "Las zonas oscuras de la evaluación ambiental que autorizó 'a ciegas' el megaproyecto de Google en Cerrillos" ["The Dark Areas of the Environmental Assessment that 'Blindly' Authorized Google's Megaproject in Cerrillos"]. CIPER. https://www.ciperchile.cl/2020/05/25/las-zonas-oscuras-de-la-evaluacion-ambiental-que-autorizo-a-ciegas-el-megaproyecto-de-google-en-cerrillos.

Armstrong, Matthew, Cade Dopp, and Jesse Welsh. 2020. "Design-Based Research." In *The Students' Guide to Learning Design and Research*, edited by R. Kimmons and S. Caskurlu. EdTech Books. https://edtechbooks.org/studentguide/design-based_research.

Ausloos, Jef, and Pierre Dewitte. 2018. "Shattering One-Way Mirrors – Data Subject Access Rights in Practice." *International Data Privacy Law* 8 (1): 4–28. https://doi.org/10.1093/idpl/ipy001.

Barnes, Susan B. 2006. "A Privacy Paradox: Social Networking in the United States." *First Monday*, September. https://doi.org/10.5210/fm.v11i9.1394.

Bartsch, Miriam, and Tobias Dienlin. 2016. "Control Your Facebook: An Analysis of Online Privacy Literacy." *Computers in Human Behavior* 56 (March): 147–54. https://doi.org/10.1016/j.chb.2015.11.022.

Bates, Jo, Yu Wei Lin, and Paula Goodale. 2016. "Data Journeys: Capturing the Socio-Material Constitution of Data Objects and Flows." *Big Data and Society* 3 (2): 1–12. https://doi.org/10.1177/2053951716654502.

Baudrillard, Jean. 1994. *Simulacra and Simulation*, 14th printing ed. Translated by Sheila Faria Glaser. University of Michigan Press.

Bauman, Zygmunt, and David Lyon. 2012. *Liquid Surveillance: A Conversation*. Wiley.

Beckett, Antony, and Ajit Nayak. 2008. "The Reflexive Consumer." *Marketing Theory* 8 (3): 299–317. https://doi.org/10.1177/1470593108093558.

Benjamin, Walter. 1969. "The Work of Art in the Age of Mechanical Reproduction." In *Illuminations*, edited by Hannah Arendt, 1–26. New York: Schocken Books.

Bhabha, Homi. 1992. "The World and the Home." *Social Text*, nos 31/32: 141–53. https://doi.org/10.2307/466222.

Bhargava, Rahul. 2019. "Data Literacy." In *The International Encyclopedia of Media Literacy*, 1st ed., edited by Renee Hobbs and Paul Mihailidis, 1–5. Wiley. https://doi.org/10.1002/9781118978238.ieml0049.

Bhargava, Rahul, Erica Deahl, Emmanuel Letouze, Amanda Noonan, David Sangokoya, and Natalie Shoup. 2015. "Beyond Data Literacy: Reinventing Community Engagement and Empowerment in the Age of Data." *Data-Pop Alliance*. https://datapopalliance.org/wp-content/uploads/2015/11/Beyond-Data-Literacy-2015.pdf.

Bhargava, Rahul, and Catherine D'Ignazio. 2015. "Designing Tools and Activities for Data Literacy Learners." *MIT Media Lab*, 1 April 2015. https://www.media.mit.edu/publications/designing-tools-and-activities-for-data-literacy-learners.

Bhargava, Rahul, Ricardo Kadouaki, Emily Bhargava, Guilherme Castro, and Catherine D'Ignazio. 2016. "Data Murals: Using the Arts to Build Data Literacy." *The Journal of Community Informatics* 12 (3). http://ci-journal.net/index.php/ciej/article/view/1276.

Boehner, Kirsten, William Gaver, and Andy Boucher. 2014. "Probes." In *Inventive Methods: The Happening of the Social*, edited by C. Lury and N. Wakeford, 185–201. Routledge.

Bogard, William. 1996. *The Simulation of Surveillance: Hypercontrol in Telematic Societies*, 1st ed. Cambridge University Press.

boyd, danah. 2018. "You Think You Want Media Literacy … Do You?" *Medium*, 16 March 2018. https://points.datasociety.net/you-think-you-want-media-literacy-do-you-7cad6af18ec2.

Boyne, Roy. 2000. "Post-Panopticism." *Economy and Society* 29 (2): 285–307. https://doi.org/10.1080/030851400360505.

Braidotti, Rosi. 2019. *Posthuman Knowledge*. Wiley.

Brevins, Katie, ed. 2018. "Bell Hooks and Consciousness-Raising: Argument for a Fourth Wave of Feminism." In *Mediating Misogyny*, 371–88. Springer International Publishing. https://doi.org/10.1007/978-3-319-72917-6_19.

Bridle, James. 2018. *New Dark Age: Technology and the End of the Future*. Verso Books.

Brunton, Finn, and Helen Nissenbaum. 2016. *Obfuscation: A User's Guide for Privacy and Protest*, reprint ed. The MIT Press.

Burgess, Jean, Kath Albury, Anthony McCosker, and Rowan Wilken. 2022. *Everyday Data Cultures*. John Wiley & Sons.

Calhoun, Craig. 1998. "Community without Propinquity Revisited: Communications Technology and the Transformation of the Urban Public Sphere." *Sociological Inquiry* 68 (3): 373–97. https://doi.org/10.1111/j.1475-682X.1998.tb00474.x.

Carmi, Elinor, Simeon J. Yates, Eleanor Lockley, and Alicja Pawluczuk. 2020. "Data Citizenship: Rethinking Data Literacy in the Age of Disinformation, Misinformation, and Malinformation." *Internet Policy Review* 9 (2). https://doi.org/10.14763/2020.2.1481.

Carranza Ko, Nusta. 2013. "Cementing Class Differences: Globalization in Peru." *Perspectives on Global Development and Technology* 12 (3): 411–26. https://doi.org/10.1163/15691497-12341264.

Carrillo, Arturo J., and Matías Jackson. 2022. "Follow the Leader? A Comparative Law Study of the EU's General Data Protection Regulation's Impact in Latin America." *ICL Journal* 16 (2): 177–262. https://doi.org/10.1515/icl-2021-0037.

Cepeda, Juan. 2017. "The Problem of Being in Latin America: Approaching the Latin American Ontological Sentipensar." *Journal of World Philosophies* 2 (1). https://scholarworks.iu.edu/iupjournals/index.php/jwp/article/view/919.

Certeau, Michel de. 1984. *The Practice of Everyday Life*, reprint ed. University of California Press.

Certeau, Michel de, Luce Giard, and Pierre Mayol. 1998. *The Practice of Everyday Life*, 1st ed. University of Minnesota Press.

Christl, Wolfie. 2017. "Corporate Surveillance in Everyday Life." *Cracked Labs*, June. https://crackedlabs.org/en/corporate-surveillance.

Chung, Grace, and Sara M. Grimes. 2006. "Data Mining the Kids: Surveillance and Market Research Strategies in Children's Online Games." *Canadian Journal of Communication* 30 (4): 527–48. https://doi.org/10.22230/cjc.2005v30n4a1525.

Clemence, Sara. 2022. "Black Travelers Say Home-Share Hosts Discriminate, and a New Airbnb Report Agrees." *The New York Times*, 13 December 2022. https://www.nytimes.com/2022/12/13/travel/vacation-rentals-racism.html.

Coslado, Ángel Barbas. 2012. "Educomunicación: desarrollo, enfoques y desafíos en un mundo interconectado" ["Educommunication: Development, Approaches and Challenges in an Interconnected World"]. *Foro de Educación* 10 (14): 157–75. https://www.redalyc.org/pdf/4475/447544618012.pdf.

Costanza-Chock, Sasha. 2020. *Design Justice: Community-Led Practices to Build the Worlds We Need*. The MIT Press.

Couldry, Nick. 2019. "Afterword." In *Citizen Media and Practice: Currents, Connections, Challenges*, edited by H.C. Stephansen and Emiliano Treré, 181–9. Routledge.

Couldry, Nick. 2023. "Hermeneutics for an Anti-Hermeneutic Age: What the Legacy of Jesús Martín-Barbero Means Today." *Media, Culture & Society* (December): 01634437231217176. https://doi.org/10.1177/01634437231217176.

Couldry, Nick, and Andreas Hepp. 2018. *The Mediated Construction of Reality*. John Wiley & Sons.

Couldry, Nick, and Ulises Ali Mejias. 2019. *The Costs of Connection: How Data Is Colonizing Human Life and Appropriating It for Capitalism*. Stanford University Press. https://doi.org/10.1515/9781503609754.

Courtland, Rachel. 2018. "Bias Detectives: The Researchers Striving to Make Algorithms Fair." *Nature* 558 (7710): 357–60. https://doi.org/10.1038/d41586-018-05469-3.

Deleuze, Gilles. 1992. "Postscript on the Societies of Control." *October* 59: 3–7.

Dencik, Lina. 2018. "Surveillance Realism and the Politics of Imagination: Is There No Alternative?" *Krisis* 1. https://www.proquest.com/docview/2291069053.

Dencik, Lina. 2019. "Situating Practices in Datafication – from Above and Below." In *Citizen Media and Practice*, 1st ed., edited by Hilde C. Stephansen and Emiliano Treré, 243–55. Routledge. https://doi.org/10.4324/9781351247375-18.

Dencik, Lina, Arne Hintz, Joanna Redden, and Emiliano Treré. 2022. *Data Justice*. SAGE Publications.

D'Ignazio, Catherine. 2017. "Creative Data Literacy: Bridging the Gap between the Data-Haves and Data-Have Nots." *Information Design Journal* 23 (July): 6–18. https://doi.org/10.1075/idj.23.1.03dig.

D'Ignazio, Catherine, and Rahul Bhargava. 2018. "Creative Data Literacy: A Constructionist Approach to Teaching Information Visualization." MIT Center for Civic Media. https://dspace.mit.edu/handle/1721.1/123473.

Dijck, Jose van. 2014. "Datafication, Dataism and Dataveillance: Big Data between Scientific Paradigm and Ideology." *Surveillance & Society* 12 (2): 197–208. https://doi.org/10.24908/ss.v12i2.4776.

Dodds, Leigh. 2017a. "Mapping Data Ecosystems." https://docs.google.com/document/d/1NG85_mdKfQUxNkBOzl9Vtvv7-aLoHupCTRHTj4F_vj8/edit#heading=h.3bx5223f7070.

Dodds, Leigh. 2017b. "What Is Data Asymmetry?" *Lost Boy* (blog), 24 March 2017. https://blog.ldodds.com/2017/03/24/what-is-data-asymmetry.

Dubal, Veena. 2023. "On Algorithmic Wage Discrimination." *SSRN Electronic Journal*. https://doi.org/10.2139/ssrn.4331080.

Dubberly. 2008. "An Evolving Map of Design Practice and Design Research." *Dubberly Design Office* (blog), 1 November 2008. https://www.dubberly.com/articles/an-evolving-map-of-design-practice-and-design-research.html.

Dubet, François. 1994. "The System, the Actor, and the Social Subject." Translated by Michael Jager. *Thesis Eleven* 38 (1): 16–35. https://doi.org/10.1177/072551369403800103.

Dubet, François. 1996. "¿Ocaso de la idea de sociedad?" *Revista de Sociología* 10: 7–23. https://doi.org/10.5354/0719-529X.1996.27653.

Dunne, Anthony, and Fiona Raby. 2001. *Design Noir: The Secret Life of Electronic Objects*. Springer Science & Business Media.

Ekbia, Hamid R., and Bonnie A. Nardi. 2017. *Heteromation, and Other Stories of Computing and Capitalism*. MIT Press.

Elisa Raffaghelli, Juliana. 2020. "Is Data Literacy a Catalyst of Social Justice? A Response from Nine Data Literacy Initiatives in Higher Education." *Education Sciences* 10 (9): 233. https://doi.org/10.3390/educsci10090233.

Escobar, Arturo. 2014. *Sentipensar con la tierra: nuevas lecturas sobre desarrollo, territorio y diferencia* [*Sentipensar* with the Land: New Readings on Development, Territory and Difference], 1st ed. Ediciones Unaula.

Escobar, Arturo. 2018. *Designs for the Pluriverse: Radical Interdependence, Autonomy, and the Making of Worlds*. Duke University Press.

Estrada, Marcela Suárez, and Sebastián Lehuedé. 2022. "Towards a Terrestrial Internet: Re-Imagining Digital Networks from the Ground Up." *Tapuya: Latin American Science, Technology and Society* 5 (1): 2139913. https://doi.org/10.1080/25729861.2022.2139913.

Fals Borda, Orlando. 1998. "Experiencias Teórico-Prácticas" ["Theoretical-Practical Experience"]. In *Participación Popular: Retos del Futuro* [*Popular Participation: Challenges of the Future*], 169–236. ICFES-IEPRI-Colciencias. https://biblioteca.clacso.edu.ar/clacso/se/20160308052028/10expe.pdf.

Fals-Borda, Orlando. 1987. "The Application of Participatory Action-Research in Latin America." *International Sociology* 2 (4): 329–47. https://doi.org/10.1177/026858098700200401.

Fleurence, Rachael L., Anne C. Beal, Susan E. Sheridan, Lorraine B. Johnson, and Joe V. Selby. 2014. "Patient-Powered Research Networks Aim to Improve Patient Care and Health Research." *Health Affairs* 33 (7): 1212–19. https://doi.org/10.1377/hlthaff.2014.0113.

Fourcade, Marion, and Kieran Healy. 2017. "Seeing Like a Market." *Socio-Economic Review* 15 (1): 9–29. https://doi.org/10.1093/ser/mww033.

Fraser, Nancy. 1995. "From Redistribution to Recognition? Dilemmas of Justice in a 'Post-Socialist' Age." *New Left Review* I, no. 212: 68–93.

Fraser, Nancy. 1998. "Social Justice in the Age of Identity Politics. Redistribution, Recognition, Participation." WZB discussion paper.

Freire, Paulo. 1970. "Cultural Action and Conscientization." *Harvard Educational Review* 40 (3): 452–77.

Freire, Paulo. 1985. *The Politics of Education: Culture, Power and Liberation*, 1st ed. Bergin & Garvey Publishers.

Fricker, Miranda. 2007. *Epistemic Injustice: Power and the Ethics of Knowing*. Oxford University Press.

Gaddis, Michael S. 2018. "An Introduction to Audit Studies in the Social Sciences." In *Audit Studies: Behind the Scenes with Theory, Method, and Nuance*, 3–44. Springer International Publishing.

Gandy, Oscar H. 1993. "Toward a Political Economy of Personal Information." *Critical Studies in Mass Communication* 10 (1): 70–97. https://doi.org/10.1080/15295039309366849.

Gandy, Oscar H. 2011. "The Political Economy of Personal Information." In *The Handbook of Political Economy of Communications*, 436–57. John Wiley & Sons. https://doi.org/10.1002/9781444395402.ch20.

Gandy, Oscar H. 2021. *The Panoptic Sort: A Political Economy of Personal Information*, 2nd ed. Oxford University Press.

Gaver, Bill, Tony Dunne, and Elena Pacenti. 1999. "Design: Cultural Probes." *Interactions* 6 (1): 21–9. https://doi.org/10.1145/291224.291235.

Gebre, Engida. 2022. "Conceptions and Perspectives of Data Literacy in Secondary Education." *British Journal of Educational Technology* 53 (5): 1080–95. https://doi.org/10.1111/bjet.13246.

Gee, James Paul. 2020. "The New Literacy Studies." In *The Routledge Handbook of Literacy Studies*, 35–48. Routledge.

Gergen, Kenneth J. 2014. "From Mirroring to World-Making: Research as Future Forming." *Journal for the Theory of Social Behaviour* 45 (3): 287–310.

Gezgin, Ulaş Başar. 2019. "Data Activism: Reviving, Extending and Upgrading Critical Citizenship Education and Consumer Rights Movements." *Connectist: Istanbul University Journal of Communication Sciences*, no. 56 (June): 67–86.

Girish, Shefali. 2022. "Exploring the Value of Adding a Data Layer to Cooperatives: Megha Farmer Cooperative Case Study." *Aapti Institute* (blog), 2 November 2022. https://aapti.medium.com/exploring-the-value-of-adding-a-data-layer-to-cooperatives-megha-farmer-cooperative-case-study-1c4fcfd08635.

Government of Canada, Innovation. 2021. "Tri-Agency Research Data Management Policy." Innovation, Science and Economic Development Canada, 14 March 2021. https://science.gc.ca/site/science/en/interagency-research-funding/

policies-and-guidelines/research-data-management/tri-agency-research-data-management-policy.

Grimes, Sara M., and Leslie Regan Shade. 2005. "Neopian Economics of Play: Children's Cyberpets and Online Communities as Immersive Advertising in NeoPets.Com." *International Journal of Media & Cultural Politics* 1 (2): 181–98. https://doi.org/10.1386/macp.1.2.181/1.

Guerrero Osorio, Arturo. 2019. "Comunalidad." In *Pluriverse: A Post-Development Dictionary*, 130–33. Tulika Books. https://www.ehu.eus/documents/6902252/12061123/Ashish+Kothari+et+al-Pluriverse+A+Post-Development+Dictionary-2019.pdf/c9f05ea0-d2e7-8874-d91c-09d11a4578a2.

Gutiérrez, Miren. 2019. "Participation in a Datafied Environment: Questions about Data Literacy." *Comunicação e Sociedade*, no. 36 (December): 37–55.

Haggerty, Kevin D., and Richard V. Ericson. 2000. "The Surveillant Assemblage." *The British Journal of Sociology* 51 (4): 605–22. https://doi.org/10.1080/00071310020015280.

Han, Byung-Chul. 2017. *Psychopolitics: Neoliberalism and New Technologies of Power*. Translated by Erik Butler. Verso.

Hare, Stephanie. 2019. "These New Rules Were Meant to Protect Our Privacy. They Don't Work." *The Observer*, 10 November 2019. https://www.theguardian.com/commentisfree/2019/nov/10/these-new-rules-were-meant-to-protect-our-privacy-they-dont-work.

Hedayati, Mona. 2025. "Affective Frequencies: From Body to Data-Made-Sound." *Somatechnics* 14 (3): 140–9. https://doi.org/10.3366/soma.2025.0453.

Hilts, Andrew, and Christoper Parsons. 2014. "Right to Information in Canada – Drawing Analogue Law into a Digital Present." *The Winston Report*. https://ssrn.com/abstract=2504109.

hooks, bell. 1986. "Sisterhood: Political Solidarity between Women." *Feminist Review* 23 (1): 125–38. https://doi.org/10.1057/fr.1986.25.

hooks, bell. 2014. *Feminism Is for Everybody: Passionate Politics*, 2nd ed. Routledge.

Horst, Rachel. 2023. "Imagining Difference: Technological Posthumanist Methods for Arts-Based Futures Literacies Research." *Digital Culture & Education* 14 (5): 214–42.

Horst, Rachel, and Derek Gladwin. 2024. "Multiple Futures Literacies: An

Interdisciplinary Review." *Journal of Curriculum and Pedagogy* 21 (1): 42–64. https://doi.org/10.1080/15505170.2022.2094510.

Houser, Nathan. 1983. "Peirce's General Taxonomy of Consciousness." *Transactions of the Charles S. Peirce Society* 19 (4): 331–59.

Johannessen, Leon Karlsen. 2017. *The Young Designer's Guide to Speculative and Critical Design. Norwegian University of Science and Technology.* https://www.ntnu.edu/documents/139799/1279149990/16+TPD4505.leon.johannessen.pdf/1c9221a2-2f1b-42fe-ba1f-24bb681beocd.

Johnson, Genevieve Marie. 2007. "Functional Internet Literacy: Required Cognitive Skills with Implications for Instruction." *E-Learning and Digital Media* 4 (4): 433–41. https://doi.org/10.2304/elea.2007.4.4.433.

Kan, Michael. 2024. "FTC Wants Answers on Companies Using Your Data for 'Surveillance Pricing.'" PCMAG, 23 July 2024. https://www.pcmag.com/news/ftc-wants-answers-on-companies-using-your-data-for-surveillance-pricing.

Kaplún, Mario. 1998. *Una pedagogía de la comunicación* [*A Pedagogy of Communication*]. Ediciones de la Torre.

Kaye, Danielle. 2024. "Landlords Used Software to Set Rents. Then Came the Lawsuits." *The New York Times*, 19 July 2024. https://www.nytimes.com/2024/07/19/business/economy/rent-prices-realpage-lawsuit.html.

Kear, Mark. 2017. "Playing the Credit Score Game: Algorithms, 'Positive' Data and the Personification of Financial Objects." *Economy and Society* 46 (3–4): 346–68. https://doi.org/10.1080/03085147.2017.1412642.

Kennedy, H. 2018. "Living with Data: Aligning Data Studies and Data Activism Through a Focus on Everyday Experiences of Datafication." *Krisis: Journal for Contemporary Philosophy* 2018 (1): 18–30.

Kitchin, Rob. 2014. *The Data Revolution: Big Data, Open Data, Data Infrastructures & Their Consequences.* SAGE Publications. https://doi.org/10.4135/9781473909472.

Kitchin, Rob. 2022. *The Data Revolution: A Critical Analysis of Big Data, Open Data & Data Infrastructures*, 2nd ed. Sage Publications.

Kitchin, Rob, and Tracey P. Lauriault. 2018. "Towards Critical Data Studies: Charting and Unpacking Data Assemblages and Their Work." In *Thinking Big Data in Geography: New Regimes, New*

Research, edited by J. Thatcher, A. Shears, and Josef Eckert, 3–10. University of Nebraska Press.

Kukutai, Tahu, and John Taylor. 2016. *Indigenous Data Sovereignty*. ANU Press. https://doi.org/10.22459/CAEPR38.11.2016.

Lameras, Petros, Sylvester Arnab, and Mark Lewis. 2024. "From Data Abundance to Informed Citizenship: 15th International Conference on Interactive Mobile Communication Technologies and Learning." In *Smart Mobile Communication & Artificial Intelligence. IMCL 2023. Lecture Notes in Networks and Systems*, edited by Michael E. Auer and Thrasyvoulos Tsiatsos, 937: 277–86. Springer.

Latour, Bruno. 2007. "Beware, Your Imagination Leaves Digital Traces." *Times Higher Literary Supplement*. http://www.bruno-latour.fr/sites/default/files/P-129-THES-GB.pdf.

Laurie, Nina, and Alastair Bonnett. 2002. "Adjusting to Equity: The Contradictions of Neoliberalism and the Search for Racial Equality in Peru." *Antipode* 34 (1): 28–53. https://doi.org/10.1111/1467-8330.00225.

Law, John. 2004. *After Method: Mess in Social Science Research*. Routledge.

Lehuedé, Sebastián. 2022. "Territories of Data: Ontological Divergences in the Growth of Data Infrastructure." *Tapuya: Latin American Science, Technology and Society* 5 (1): 2035936. https://doi.org/10.1080/25729861.2022.2035936.

Lewis, T., S.P. Gangadharan, M. Saba, and T. Petty. 2018. *Our Data Bodies: Digital Defense Playbook*. Our Data Bodies. https://www.odbproject.org/wp-content/uploads/2019/03/ODB_DDP_HighRes_Single.pdf.

Lima Pinel, Maria de Fátima de, José Paulo Cosenza, and Fernando Llena Macarulla. 2015. "La auditoría social como mecanismo de control de la responsabilidad social de las empresas: la metodología de Theodore J. Kreps" ["Social Auditing as a Mechanism for Monitoring Corporate Social Responsibility: The Methodology of Theodore J. Kreps"]. *Contabilidad y Negocios* 10 (19): 85–100. https://doi.org/10.18800/contabilidad.201501.005.

LinkedIn. 2024 "LinkedIn Talent Insights." Accessed 18 December 2024. https://business.linkedin.com/talent-solutions/talent-insights.

López, Joan, and Fundación Karisma. 2020. "Experimentando con la pobreza: El Sisbén y los proyectos de analítica de datos en Colombia" ["Experimenting with Poverty: Sisbén and Data Analytics Projects in Colombia"]. Fundación Karisma. https://doi.org/10.13140/RG.2.2.19489.15207.

Lupton, Deborah. 2021. "'Not the Real Me': Social Imaginaries of Personal Data Profiling." *Cultural Sociology* 15 (1): 3–21. https://doi.org/10.1177/1749975520939779.

Lust, Jan. 2019. "Neoliberalism and the Socialist Left in Peru: An Epochal Change." In *Capitalism, Class and Revolution in Peru, 1980-2016*, edited by Jan Lust, 11–24. Springer International Publishing. https://doi.org/10.1007/978-3-319-91403-9_2.

Lyon, David. 2002. "Surveillance as Social Sorting: Computer Codes and Mobile Bodies." In *Surveillance as Social Sorting: Privacy, Risk, and Digital Discrimination*. Routledge.

Lyon, David. 2007. *Surveillance Studies: An Overview*, 13–30. Polity.

Macedo, Donaldo. 2022. "The Centrality of Conscientization in Critical Pedagogy." In *Handbook of Critical Approaches to Politics and Policy of Education*, 23–31. Routledge.

Mann, Steve, Jason Nolan, and Barry Wellman. 2003. "Sousveillance: Inventing and Using Wearable Computing Devices for Data Collection in Surveillance Environments." *Surveillance & Society* 1 (3): 331–55. https://doi.org/10.24908/ss.v1i3.3344.

Mansell, Robin, and W.E. Steinmueller. 2020. *Advanced Introduction to Platform Economics*. Edward Elgar Publishing Limited.

Markham, Annette. 2019. "Critical Pedagogy as a Response to Datafication." *Qualitative Inquiry* 25 (8): 754–60. https://doi.org/10.1177/1077800418809470.

Markham, Annette. 2020. "Taking Data Literacy to the Streets: Critical Pedagogy in the Public Sphere." *Qualitative Inquiry* 26 (2). https://doi.org/10.1177/1077800419859024.

Markham, Annette. 2021a. "The Limits of the Imaginary: Challenges to Intervening in Future Speculations of Memory, Data, and Algorithms." *New Media & Society* 23 (2): 382–405. https://doi.org/10.1177/1461444820929322.

Markham, Annette. 2021b. "The Limits of the Imaginary: Challenges to Intervening in Future Speculations of Memory, Data, and Algorithms." *New Media & Society* 23 (2): 382–405. https://doi.org/10.1177/1461444820929322.

Martin, Aaron K., Rosamunde E. van Brakel, and Daniel J. Bernhard.

2009. "Understanding Resistance to Digital Surveillance: Towards a Multi-Disciplinary, Multi-Actor Framework." *Surveillance & Society* 6 (3): 213–32. https://doi.org/10.24908/ss.v6i3.3282.

Martin, Allan. 2008. "Digital Literacy and the 'Digital Society.'" In *Digital Literacies: Concepts, Policies and Practices*, edited by Colin Lankshear and Michele Knobel, 30: 151–76. Peter Lang. https://researchonline.jcu.edu.au/27788/1/27788_Lankshear_and_Knobel_2008.pdf.

Martin Valdez, Stuard. 2020. "Encuesta de Transformación Digital" ["Digital Transformation Survey"]. Real Time Management and the Universidad de Piura. https://www.rtm.com.pe/wp-content/uploads/2021/01/Encuesta-TD-2020.pdf.

Martín-Barbero, Jesús. 1993. *Communication, Culture and Hegemony: From the Media to Mediations*. Sage.

Martín-Barbero, Jesús. 2011. "La Pertenencia En El Horizonte de Las Nuevas Tecnologías y de La Sociedad de La Comunicación" ["Belonging in the Horizon of New Technologies and the Communication Society"]. In *Sentido de Pertenencia En Sociedades Fragmentadas*, edited by Martín Hopenhayn and Ana Sojo, 105–26. Siglo XXI.

Martín-Barbero, Jesús, and Omar Rincón. 2019. "Mapa Insomne 2017: Ensayos Sobre el Sensorium Contemporáneo. Un Mapa Para Investigar la Mutación Cultural" ["Sleepless Map 2017: Essays on the Contemporary Sensorium. A Map for Investigating Cultural Mutation"]. In *Un Nuevo Mapa Para Investigar la Mutación Cultural: Diálogo con la Propuesta de Jesús Martín-Barbero* [*A New Map for Researching Cultural Mutation: Dialogue with Jesús Martín-Barbero's Proposal*], edited by Omar Rincón, 17–23. Ediciones Ciespal.

Martinez, Daniel E. 2011. "Beyond Disciplinary Enclosures: Management Control in the Society of Control." *Critical Perspectives on Accounting* 22 (2), "Norman Macintosh and Critical Accounting Research: A Festschrift": 200–11. https://doi.org/10.1016/j.cpa.2010.06.016.

Martínez Luna, Jaime. 2015. "Conocimiento & Comunalidad" ["Knowledge & Communality"]. *Bajo el Volcan* 15 (23). https://www.redalyc.org/pdf/286/28643473006.pdf.

Masur, Philipp K. 2020. "How Online Privacy Literacy Supports Self-Data Protection and Self-Determination in the Age of Information." *Media and Communication* 8 (2): 258–69. https://doi.org/10.17645/mac.v8i2.2855.

Mattioli, Michael. 2017. "The Data-Pooling Problem." *Berkeley Technology Law Journal* 32 (1): 179–236.

Mayer-Schönberger, Viktor, and Kenneth Cukier. 2013. *Big Data: A Revolution That Will Transform How We Live, Work, and Think.* Mariner Books.

McBride, Sarah. 2014. "Uber's Fight of California Data-Sharing Rule Highlights Its Bumpy Road." *Reuters*, 19 December 2014. https://www.reuters.com/article/us-uber-california-data-idUSKBN0JX01320141219.

Mejias, Ulises Ali. 2023. "AI and the Tyranny of the Data Commons." *Al Jazeera*, 19 August 2023. https://www.aljazeera.com/opinions/2023/8/19/ai-and-the-tyranny-of-the-data-commons.

Mignolo, Walter D. 2005. *The Idea of Latin America.* Wiley-Blackwell Publishing.

Milan, Stefania. 2019. "Acting on Data(Fication)." In *Citizen Media and Practice*, 1st ed., edited by Hilde C. Stephansen and Emiliano Treré, 212–26. Routledge. https://doi.org/10.4324/9781351247375-16.

Milan, Stefania, and Miren Gutiérrez. 2015. "Citizens' Media Meets Big Data: The Emergence of Data Activism." *Mediaciones* 11 (14): 120–33. https://doi.org/10.26620/uniminuto.mediaciones.11.14.2015.120-133.

Milan, Stefania, and Emiliano Treré. 2019. "Big Data from the South(s): Beyond Data Universalism." *Television & New Media* 20 (4): 319–35. https://doi.org/10.1177/1527476419837739.

Milan, Stefania, and Lonneke van der Velden. 2016. "The Alternative Epistemologies of Data Activism." *Digital Culture & Society* 2 (2): 57–74. https://doi.org/10.14361/dcs-2016-0205.

Miller, Peter, and Nikolas Rose. 1997. "Mobilizing the Consumer: Assembling the Subject of Consumption." *Theory, Culture & Society* 14 (1): 1–36. https://doi.org/10.1177/026327697014001001.

Morales, Esteban, and Katherine Reilly. 2023. "The Unhomed Data Subject: Negotiating Datafication in Latin America." *Information, Communication & Society* 26 (12): 2457–71. https://doi.org/10.1080/1369118X.2023.2250436.

Morales, Susana. 2009. "La apropiación de TIC: una perspectiva" ["ICT Appropriation: A Perspective"]. In *Los jóvenes y las TIC. Apropiación y uso en educación* [*Young People and ICT. Appropriation and Use in Education*], edited by Susana Morales and Maria Ines Loyola, 36–45. Argentina.

Morales, Susana. 2018. "La apropiación de tecnologías. Ideas para un paradigma en construcción" ["The Appropriation of Technologies. Ideas for a Paradigm under Construction"]. In *Acerca de la apropiación de tecnologías: Teoría, estudios y debates* [*About the Appropriation of Technologies: Theory, Studies and Debates*], edited by Silvia Lago Martínez, Ayelén Álvarez, Martín Ariel Gendler, and Anahí Méndez, 23–34. Buenos Aires: Ediciones del gato gris. https://www.aacademica.org/anahi.mendez/77.pdf

More Perfect Union, dir. 2024. "We Put 7 Uber Drivers in One Room. What We Found Will Shock You." Accessed 27 August 2024. https://www.youtube.com/watch?v=DLEkeXtr-zo.

Mundie, Craig. 2014. "Privacy Pragmatism: Focus on Data Use, Not Data Collection." *Foreign Affairs* 93 (2): 28–38.

Najibi, Alex. 2020. "Racial Discrimination in Face Recognition Technology." *Science in the News* (blog), 24 October 2020. https://sitn.hms.harvard.edu/flash/2020/racial-discrimination-in-face-recognition-technology.

Nebraska Association of Resources Districts. 2019. "Twin Platte NRD Receives $1.6M Grant for Water Data Program." 19 December 2019. https://www.nrdnet.org/news/12-19-2019/twin-platte-nrd-receives-16m-grant-water-data-program.

Neff, Gina, Anissa Tanweer, Brittany Fiore-Gartland, and Laura Osburn. 2017. "Critique and Contribute: A Practice-Based Framework for Improving Critical Data Studies and Data Science." *Big Data* 5 (2): 85–97. https://doi.org/10.1089/big.2016.0050.

O'Reilly, Tim. 2005. "What Is Web 2.0: Design Patterns and Business Models for the Next Generation of Software." *O'Reilly*, 30 September 2005. https://www.oreilly.com/pub/a/web2/archive/what-is-web-20.html.

Oropeza Alvarez, Daliri. 2024. "EZLN celebra 30 años con alegría en común" ["EZLN Celebrates 30 Years with Joy in Common"]. *Pie de Página*, 3 January 2024. https://piedepagina.mx/ezln-celebra-30-anos-con-alegria-en-comun.

Pangrazio, Luci, and Neil Selwyn. 2019. "'Personal Data Literacies': A Critical Literacies Approach to Enhancing Understandings of Personal Digital Data." *New Media & Society* 21 (2): 419–37. https://doi.org/10.1177/1461444818799523.

Pangrazio, Luci, and Neil Selwyn. 2023. *Critical Data Literacies: Rethinking Data and Everyday Life*. The MIT Press.

Park, Yong Jin. 2013. "Digital Literacy and Privacy Behavior Online." *Communication Research* 40 (2): 215–36. https://doi.org/10.1177/0093650211418338.

Parry, Benita. 1994. "Signs of Our Times: Discussion of Homi Bhabha's *The Location of Culture*." *Third Text*, September. https://doi.org/10.1080/09528829408576499.

Patient-Led Research Collaborative. n.d. "Frequently Asked Questions." Accessed 23 September 2024. https://patientresearchcovid19.com/faq.

Pedersen, Katie, Greg Sadler, and Virginia Smart. 2017. "How Companies Use Personal Data to Charge Different People Different Prices for the Same Product." *CBC News*, 24 November 2017. https://www.cbc.ca/news/business/marketplace-online-prices-profiles-1.4414240.

Peirce, Charles Sanders. 1955. "Logic as Semiotic: The Theory of Signs." In *Philosophical Writings of Peirce*, edited by J. Buchler, 98–119. Dover Publications.

Pink, Sarah, Debora Lanzeni, and Heather Horst. 2018. "Data Anxieties: Finding Trust in Everyday Digital Mess." *Big Data & Society* 5 (1): 2053951718756685. https://doi.org/10.1177/2053951718756685.

Poster, Mark. 1990. *The Mode of Information: Poststructuralism and Social Context*. Polity Press.

Power, Michael. 1999. *The Audit Society: Rituals of Verification*. Oxford University Press.

Press, Larry. 1997. "Tracking the Global Diffusion on the Internet." *Communications of the ACM* 40 (11): 11–17. https://doi.org/10.1145/265684.265686.

Pybus, Jennifer, Mark Coté, and Tobias Blanke. 2015. "Hacking the Social Life of Big Data." *Big Data & Society* 2 (2): 2053951715616649. https://doi.org/10.1177/2053951715616649.

Raffaghelli, Juliana E. 2019. "'Datificación' y Educación Superior: Hacia la construcción de un marco para la alfabetización en datos del profesorado universitario" ["'Datification' and Higher Education:

Toward Building a Framework for Data Literacy for University Faculty"]. *Revista Interamericana de Investigación, Educación y Pedagogía,* RIIEP 13 (1): 177–205. https://doi.org/10.15332/25005421/5466.

Ramalho, Tania. 2022. "Paulo Freire, Communication, and Conscientization for Liberation." In *The Handbook of Global Interventions in Communication Theory*, 1st ed., by Yoshitaka Miike and Jing Yin, 269–83. Routledge. https://doi.org/10.4324/9781003043348-20.

Reilly, Katherine. 2020. "The Challenge of Decolonizing Big Data through Citizen Data Audits." *BigDataSur* (blog), 30 April 2020. https://data-activism.net/2020/04/bigdatasur-the-challenge-of-decolonizing-big-data-through-citizen-data-audits-1-3.

Reilly, Katherine. 2022. "Son Mis Datos: Building Personal Data Literacies through Citizen Data Audits." In *International Conference on Information & Communication Technologies and Development 2022*, 1–9. Seattle: ACM. https://doi.org/10.1145/3572334.3572387.

Reilly, Katherine. 2024. "Communicative Sovereignty." In *Encyclopedia of Political Communication*, edited by Max Gromping, Dominique Wirz, and Alessandro Nai, 249–52. Edward Elgar Publishers.

Reilly, Katherine, Belén Febres-Cordero, Esteban Morales, Denisse Albornoz, Miguel Morachimo, and Marieliv Flores. 2020. "Son Mis Datos: Guía Metodológica" ["It's My Data: Methodological Guide"]. *Hiperderecho.* https://hiperderecho.org/wp-content/uploads/2020/07/SonMisDatos.pdf.

Reilly, Katherine, and Marieliv Flores. Forthcoming. "Contextualizing Datafication in Peru: Insights from a Citizen Data Literacy Project." In *Putting Data Justice in Context*, edited by Tone Walford, Sandeep Mertia, and Catarina Morawska-Vianna. UCL Press.

Reilly, Katherine, and Esteban Morales. 2023. "Citizen Data Audits in the Contemporary Sensorium." *International Journal of Communication* 17: 3582–99. https://doi.org/1932–8036/20230005.

Reilly, Katherine, and Esteban Morales. Forthcoming. "The Unhomed Data Subject." In *Handbook of Critical Data Studies*, edited by Ulises A. Mejias, Jasmine McNealy, and Milagros Miceli. De Gruyter.

Reilly, Katherine, and Maria Julia Morales. 2024. "The Challenge of Addressing Subjectivities through Participatory Action Research on Datafication." In *Collaborative Research in the Datafied Society: Methods and Practices for Investigation and Intervention*, edited by

Mirko Schaefer, Tracey Lauriault, and Karin van Es. Amsterdam University Press.

Ricaurte, Paola. 2019. "Data Epistemologies, The Coloniality of Power, and Resistance." *Television & New Media* 20 (4): 350–65. https://doi.org/10.1177/1527476419831640.

Ricaurte, Paola. 2022. "Ethics for the Majority World: AI and the Question of Violence at Scale." *Media, Culture & Society* 44 (4): 726–45. https://doi.org/10.1177/01634437221099612.

Rincón, Omar. 2018. "Mutaciones bastardas de la comunicación" ["Bastard Mutations of Communication"]. *MATRIZES* 12 (1): 65. https://doi.org/10.11606/issn.1982-8160.v12i1p65-78.

Rivoir, Ana, and Javier Landinelli. 2017. "Country Overview Uruguay." IT for Change. https://opendocs.ids.ac.uk/opendocs/handle/20.500.12413/13333.

Roa, Luisa, Alejandro Correa-Bahnsen, Gabriel Suarez, Fernando Cortés-Tejada, María A. Luque, and Cristián Bravo. 2021. "Super-App Behavioral Patterns in Credit Risk Models: Financial, Statistical and Regulatory Implications." arXiv.org, 9 May 2021. https://doi.org/10.1016/j.eswa.2020.114486.

Roberts, Kenneth M. 1995. "Neoliberalism and the Transformation of Populism in Latin America: The Peruvian Case." *World Politics* 48 (1): 82–116.

Rosenberg, Daniel. 2013. "Data before the Fact." In *Raw Data Is an Oxymoron*, edited by Lisa Gitelman, 15–40. MIT Press.

Rowley, Jennifer. 2007. "The Wisdom Hierarchy: Representations of the DIKW Hierarchy." *Journal of Information Science* 33 (2): 163–80. https://doi.org/10.1177/0165551506070706.

Sander, Ina. 2020. "What Is Critical Big Data Literacy and How Can It Be Implemented?" *Internet Policy Review* 9 (2). https://policyreview.info/articles/analysis/what-critical-big-data-literacy-and-how-can-it-be-implemented.

Sastre Domínguez, Paz, and Ángel Juan Gordo López. 2019. "Data Activism versus Algorithmic Control. New Governance Models, Old Asymmetries." *Docta Complutense*. https://hdl.handle.net/20.500.14352/13860.

Scassa, Teresa. 2018. "Supreme Court of Canada Decision Has Relevance for Addressing Bias in Algorithmic Decision-Making." *Teresa Scassa* (blog), 14 June 2018. https://www.teresascassa.ca/index.

php?option=com_k2&view=item&id=278:supreme-court-of-canada-decision-has-relevance-for-addressing-bias-in-algorithmic-decision-making&Itemid=80.

Schwan, Hannah, Jonas Arndt, and Marian Dörk. 2022. "Disclosure as a Critical-Feminist Design Practice for Web-Based Data Stories." *First Monday*, November. https://doi.org/10.5210/fm.v27i11.12712.

Scott, James C. 1998. *Seeing Like a State: How Certain Schemes to Improve the Human Condition Have Failed.* Yale University Press.

Sefton-Green, Julian, and Luci Pangrazio. 2021. "The Death of the Educative Subject? The Limits of Criticality under Datafication." *Educational Philosophy and Theory* 54 (12): 2072–81. https://doi.org/10.1080/00131857.2021.1978072.

Segura, María Soledad, and Silvio Waisbord. 2019. "Between Data Capitalism and Data Citizenship." *Television & New Media* 20 (4): 412–19. https://doi.org/10.1177/1527476419834519.

Shreiner, Tamara L., and Wayne Journell. 2024. *Teaching Data Literacy in Social Studies: Principles and Practices to Support Historical Thinking and Civic Engagement.* Teachers College Press.

Siles, Ignacio. 2023. "The Mutual Domestication of Users and Algorithms: The Case of Netflix." In *The Routledge Handbook of Media and Technology Domestication*, edited by Maren Hartmann, 235–48. Routledge.

Siles, Ignacio, Edgar Gómez-Cruz, and Paola Ricaurte. 2023. "Fluid Agency in Relation to Algorithms: Tensions, Mediations, and Transversalities." *Convergence* (May): 13548565231174586. https://doi.org/10.1177/13548565231174586.

Silverstone, Roger. 1989. "Let Us Then Return to the Murmuring of Everyday Practices: A Note on Michel de Certeau, Television and Everyday Life." *Theory, Culture & Society* 6, no. 1: 77–94.

Silverstone, Roger. 1994. *Television and Everyday Life.* Routledge.

Singh, Ranjit, and Steven Jackson. 2021. "Seeing Like an Infrastructure: Low-Resolution Citizens and the Aadhaar Identification Project." *Proceedings of the ACM on Human-Computer Interaction* 5 (CSCW2): 1–26. https://doi.org/10.1145/3476056.

Smith, Karl. 2012. "From Dividual and Individual Selves to Porous Subjects." *The Australian Journal of Anthropology* 23 (1): 50–64. https://doi.org/10.1111/j.1757-6547.2012.00167.x.

Soden, Robert, David Lallemant, Perrine Hamel, and Karen Barns. 2021. "Becoming Interdisciplinary: Fostering Critical Engagement With Disaster Data." *Proceedings of the ACM on Human-Computer Interaction* 5 (CSCW1): 168: 1–27. https://doi.org/10.1145/3449242.

Southworth, Cindy, Shawndell Dawson, Cynthia Fraser, and Sarah Tucker. 2005. "A High-Tech Twist on Abuse: Technology, Intimate Partner Stalking, and Advocacy." *National Network to End Domestic Violence Fund.* https://www.acesdv.org/wp-content/uploads/2014/06/NNEDV_HighTechTwist_PaperAndApxA_English08.pdf.

Square Support. n.d. "Get Started with Customer Information Collection." https://squareup.com/help/ca/en/article/5976-get-started-with-email-collection-tools.

Srnicek, Nick. 2017. "The Challenges of Platform Capitalism: Understanding the Logic of a New Business Model." *Juncture* 23 (4): 254–57. https://doi.org/10.1111/newe.12023.

Star, Susan Leigh, and Geoffrey C. Bowker. 2007. "Enacting Silence: Residual Categories as a Challenge for Ethics, Information Systems, and Communication." *Ethics and Information Technology* 9 (4): 273–80. https://doi.org/10.1007/s10676-007-9141-7.

Street, Brian. 2003. "What's 'New' in New Literacy Studies? Critical Approaches to Literacy in Theory and Practice." *Current Issues in Comparative Education* 5 (2). https://doi.org/10.52214/cice.v5i2.11369.

Sursiendo. 2022. "Comunalidad Digital: una aproximación desde la ética permacultural" ["Digital Communality: An Approach from Permacultural Ethics"]. Chiapas, Mexico: Sursiendo. https://sursiendo.org/2022/04/comunalidad-digital-una-aproximacion-desde-la-etica-permacultural.

Sursiendo. n.d. "Somos / About Us." Accessed 23 September 2024. https://sursiendo.org/quienes-somos.

Taylor, Humphrey. 2003. "Most People Are 'Privacy Pragmatists' Who, While Concerned about Privacy, Will Sometimes Trade It Off for Other Benefits." Harris Interactive. https://www.researchgate.net/publication/268342423_Most_People_Are_Privacy_Pragmatists_Who_While_Concerned_about_Privacy_Will_Sometimes_Trade_It_Off_for_Other_Benefits.

Taylor, Linnet, and Dennis Broeders. 2015. "In the Name of Development: Power, Profit and the Datafication of the Global

South." *Geoforum* 64 (August): 229–37. https://doi.org/10.1016/j.geoforum.2015.07.002.

TEDIC. 2023. "Por la Defensa de los Datos Personales en Paraguay" ["For the Defense of Personal Data in Paraguay"]. Tecnologia & Derechos Humanos (TEDIC).

Thatcher, Jim, David O'Sullivan, and Dillon Mahmoudi. 2016. "Data Colonialism through Accumulation by Dispossession: New Metaphors for Daily Data." *Environment and Planning D: Society and Space* 34 (6): 990–1006. https://doi.org/10.1177/0263775816633195.

The Privacy Settings Collective. 2023. "Reading the Data Subject." *Post45* (blog), 8 December 2023. https://post45.org/2023/12/reading-the-data-subject.

Tironi, Martín, and Matías Valderrama. 2021. "Descolonizando los sistemas algorítmicos: diseño crítico para la problematización de algoritmos y datos digitales desde el Sur" ["Decolonizing Algorithmic Systems: Critical Design for the Problematization of Algorithms and Digital Data from the South"]. *Palabra Clave* 24 (3): 1–33. https://doi.org/10.5294/pacla.2021.24.3.2.

Trepte, Sabine, Doris Teutsch, Philipp K. Masur, C. Eichler, Mona Fischer, Alisa Hennhöfer, and Fabienne Lind. 2015. "Do People Know About Privacy and Data Protection Strategies? Towards the 'Online Privacy Literacy Scale' (OPLIS)." In *Reforming European Data Protection Law*, edited by Serge Gutwirth, Ronald Leenes, and Paul de Hert, 333–65. Springer. https://doi.org/10.1007/978-94-017-9385-8.

Turow, Joseph. 2003. "Americans and Online Privacy: The System Is Broken." Annenberg Public Policy Center of the University of Pennsylvania. https://core.ac.uk/download/pdf/219378111.pdf.

Turow, Joseph, Lauren Feldman, and Kimberly Meltzer. 2005. "Open to Exploitation: American Shoppers Online and Offline." Annenberg Public Policy Center of the University of Pennsylvania. https://core.ac.uk/reader/219378110.

Turow, Joseph, and Michael Hennessy. 2007. "Internet Privacy and Institutional Trust: Insights from a National Survey." *New Media & Society* 9 (2): 300–18. https://doi.org/10.1177/1461444807072219.

Tygel, Alan, and Rosana Kirsch. 2015. "Contributions of Paulo Freire for a Critical Data Literacy." *The Journal of Community Informatics* 123 (3): 108–21.

UTEC and Everis. 2021. "Transformatción Digital en el Perú: Testimonios y Experiencieas de los Principales Líderes Empresariales del Perú." Universidad de Ingeniería & Tecnología and Everis. https://asep.pe/wp-content/uploads/2021/11/Transformacio%CC%81n-digital-en-el-Peru%CC%81-1.pdf.

Vial, Gregory. 2019. "Understanding Digital Transformation: A Review and a Research Agenda." *The Journal of Strategic Information Systems* 28 (2), "SI: Review Issue": 118–44. https://doi.org/10.1016/j.jsis.2019.01.003.

Villasante, Tomás R. 1999. "Cuatro Redes Para Hacer Transformaciones Sustentables" ["Four Networks for Making Sustainable Transformations"]. *Política y Sociedad* 31: 37–54.

Wilkinson, Mark D., Michel Dumontier, IJsbrand Jan Aalbersberg, Gabrielle Appleton, Myles Axton, Arie Baak, Niklas Blomberg, et al. 2016. "The FAIR Guiding Principles for Scientific Data Management and Stewardship." *Scientific Data* 3 (1): 160018. https://doi.org/10.1038/sdata.2016.18.

Williams, Emelia. 2023. "Collaborative Environmental Data Stewardship: Opportunities for Ecosystem Building and Project Application." *Open Environmental Data Project*. https://drive.google.com/file/d/1ed9QINM5L6MmJnzWApZ3Ptbo8o91onmp/view.

Williams, Raymond. 1961. *The Long Revolution*, reprint ed. Parthian Books.

Young, Andrew, Stefaan G. Verhulst, Nadiya Safonova, and Andrew Zahuranec. 2020. "The Data Assembly: Responsible Data Re-Use Framework." *The Gov Lab*. https://thedataassembly.org/files/nyc-data-assembly-report.pdf.

Zuboff, Shoshana. 2019. *The Age of Surveillance Capitalism: The Fight for a Human Future at the New Frontier of Power*, main ed. Profile Books.

Zwick, Detlev, Samuel K. Bonsu, and Aron Darmody. 2008. "Putting Consumers to Work: 'Co-Creation' and New Marketing Govern-Mentality." *Journal of Consumer Culture* 8 (2): 163–96. https://doi.org/10.1177/1469540508090089.

Zwick, Detlev, and Alan Bradshaw. 2018. "Biopolitical Marketing and the Commodification of Social Contexts." In *The Routledge Companion to Critical Marketing*, edited by Mark Tadajewski, Matthew Higgins, Janice Denegri-Knott and Rohit Varman, 430–38. Routledge.

AUTHORS AND CONTRIBUTORS

AUTHORS

KATHERINE M.A. REILLY is an associate professor in the School of Communication at Simon Fraser University in Vancouver, Canada. Her PhD in political science from the University of Toronto focused on digital technology in Latin America. She is the co-editor of *Open Development: Networked Innovations in International Development* (with Matthew Smith; MIT Press, 2014), and lead producer of *Dialogando Sobre Datos: Un Audiolibro Sobre el Colonialismo de Datos* [Talking about Data: An Audiobook about Data Colonialism] (with Esteban Morales and Ana Girón; Tierra Común, 2025). She is also the co-editor of special editions of the journal *Somatechnics* (with Gillian Russell and Rachel Horst) and the *Journal of Developing Societies* (with Anil Hira). She has published extensively in prestigious journals including *Information, Communication & Society*, *Media and Communication*, the *International Journal of Communication*, and *Internet Policy Review*.

ESTEBAN MORALES is an assistant professor in media and digital cultures at the University of Groningen in the Netherlands. He holds a PhD in language and literacy education from the University of British Columbia in Vancouver, Canada, and was a postdoctoral fellow with Dr Jaigris Hodson at Royal Roads University in Victoria, Canada. His research focuses on how citizens and communities learn about, resist, and appropriate digital media. He is published in *New Media and Society*, *Social Media & Society*, *Information, Communication & Society*, the *International Journal of Communication*, and *First Monday*, and is currently co-editing a special edition of *New Media & Society* on "Contextual Complexities of Violence on Digital Platforms" with Martin Lundqvist and Tom Divon.

MARIA JULIA MORALES GONZALEZ is a sociologist and holds a doctorate in educational technology from the Universitat Rovira i Virgili in Catalonia, Spain. She is an instructor with the Facultad Latino Americano de Ciencias Sociales (FLACSO) and at the University of the Republic in Montevideo, Uruguay. In addition, she is a researcher with the Observatory of Information and Communication Technologies (ObservaTIC) in the Faculty of Social Sciences at the same institution. She was the project coordinator for Project Flor del Ceibo, which researched national digital literacy policies and interventions in Uruguay, and has also done extensive research on technology use by older persons in that country.

RESEARCH COLLABORATORS AND CASE STUDY AUTHORS

CAROLINA CARRETERO is a student in the Master's in Public Policy program at the University of the Republic in Montevideo, Uruguay, and a research assistant at ObservaTIC in the Faculty of Social Sciences.

EDUARDO CARRILLO is a public policy and advocacy analyst at TEDIC in Asunción, Paraguay. He holds a master's in public administration with an emphasis in digital technology and policy from University College London (UCL).

JUAN DIEGO CASTAÑEDA is a lawyer and researcher at Fundación Karisma in Bogotá, Colombia. He works on issues related to surveillance, privacy, and digital rights.

LAURA CASTRO is an anthropologist working with Fundación Karisma in Bogotá, Colombia. She has a master's in gender and policy studies, and her work focuses on labour policy.

DIANA GARCÍA is a researcher and expert in gender and digital violence based in Asunción, Paraguay. She has collaborated on several projects with TEDIC.

JUAN CARLOS LARA is the co-executive director of Derechos Digitales in Santiago, Chile. He is a lawyer, and also received a master's in law and technology from the University of California at Berkeley.

RAMIRO LIESEGANG is a student in the Master's in Public Policy program at the University of the Republic in Montevideo, Uruguay, and a research assistant at ObservaTIC in the Faculty of Social Sciences.

MARIA JOSÉ LIRA MORÁN worked as a researcher and project coordinator for HiperDerecho in Lima, Peru. She holds a master's in digital society from the University of Edinburgh.

ANA LAURA RIVOIR is a professor of sociology at the University of the Republic in Montevideo, Uruguay, and director of the research centre ObservaTIC.

DANIELA SALAS ZUÑIGA worked as a researcher and project coordinator for HiperDerecho in Lima, Peru. She holds a degree in sociology from the Pontificia Universidad Católica del Perú.

PATRICIO VELASCO is a researcher with Derechos Digitales in Santiago, Chile. He holds a master's in media and communications and works as an instructor in the School of Communication and Journalism at Universidad Adolfo Ibáñez.

INDEX